Defying Dystopia

Defying Dystopia

Going on with the Human Journey after Technology Fails Us

Ed Ayres

Transaction Publishers
New Brunswick (U.S.A.) and London (U.K.)

This book is printed on acid-free paper that meets the American National Standard for Permanence of Paper for Printed Library Materials.

Library of Congress Catalog Number: 2015045765
ISBN: 978-1-4128-6270-7 (cloth); 978-1-4128-6323-0 (paper)
eBook: 978-1-4128-6220-2
Printed in the United States of America

Library of Congress Cataloging-in-Publication Data

Names: Ayres, Ed, author.
Title: Defying dystopia : going on with the human journey after technology fails us / Ed Ayres.
Description: New Brunswick (U.S.A.) : Transaction Publishers, [2016] | Includes bibliographical references.
Identifiers: LCCN 2015045765 (print) | LCCN 2015050240 (ebook) | ISBN 9781412862707 (hardcover : acid-free paper) | ISBN 9781412863230 (pbk. : acid-free paper) | ISBN 9781412862202 (ebook)
Subjects: LCSH: Technology--Social aspects. | Social psychlogy. Resilience | (Personality trait)
Classification: LCC T14.5 .A985 2016 (print) | LCC T14.5 (ebook) | DDC 303.48/3--dc23
LC record available at http://lccn.loc.gov/2015045765

Contents

Prologue: Waking Up

On a cool November morning in 1992, I was sitting in my office at the Worldwatch Institute in Washington, DC, sorting through the usual heap of correspondence, clippings, and notes on my desk and trying not to knock over my coffee. I didn't know then that just a few years later, along with sorting paper, I'd have to wade through the morning's avalanche of e-mail—a technology that was good to have but that also seemed at times to be the digital equivalent of the guy on the corner opening his raincoat to offer dirty pictures and cheap drugs.

My office was a few feet down the corridor from the corner office of Lester R. Brown, who had founded Worldwatch as the world's first international environmental research group. Across the street from Lester's window was the Brookings Institute, where policy experts did research for presidents and senators, and where you could often see a TV crew interviewing one of the senior associates on the lawn out front, under a great old cherry tree—a tree that, when it bloomed every spring, was one of the joys of Lester Brown's life. From my window, I could see the roof patio of Planned Parenthood's national headquarters next door, where members of the embattled organization's staff often came out for a break. Half a block away was Dupont Circle, where the iconic bookstore Second Story Books often had a sidewalk bin where you could search for old treasures. Our building, 1776 Massachusetts Avenue, was on a leafy block that housed a score of nonprofits, think tanks, and grad-school annexes where bright young people could work with eminent greybeards.

Lester had recently hired me to edit the Institute's bi-monthly magazine, *World Watch*, and that morning I was just getting focused on priorities for the next issue when our mail-room guy came in and handed me a press release. It was from the Cambridge, Massachusetts-based Union of Concerned Scientists. For a moment, when I glanced at it, I was reminded of a forlorn-looking man I'd seen standing on the sidewalk a few days before, holding a hand-lettered cardboard sign announcing the End of the World. But as I looked more closely, I saw that the press release contained a statement that had been signed by 104 Nobel Prize winners in the sciences, along with more than 1,500 other senior scientists from 71 countries. "**A Threat to Human Survival,**" read the headline. The statement read:

> We the undersigned senior members of the world's scientific community, hereby warn all humanity of what lies ahead. A great change in the stewardship of the earth and life on it is required if vast human misery is to be avoided and our global home on this planet is not to be irretrievably mutilated . . . . Humans and the natural world are on a collision course.

Among the signers were the renowned astrophysicists Carl Sagan and Stephen Hawking, evolutionary biologists Stephen Jay Gould, Edward O. Wilson, and Lynn Margulis, population biologist Paul Ehrlich, anthropologist Sarah Hrdy, economist Wassily Leontief, chemist Linus Pauling, DNA pioneer James Watson (co-discoverer of the double helix structure), and astronomer James Van Allen (for whom the Van Allen Belt around the earth is named).

For the public of the early '90s, the *World Scientists' Warning to Humanity* might have been quite a shock, even if a bit euphemistic (the wording had to have been acceptable to more than 1,500 strong-minded scientists in multiple disciplines, all of whom must have known their reputations could be at stake). It didn't use inflammatory language, like "possible end of civilization," but the implications were unmistakable. To me, it seemed quite arguably one of the most important documents ever issued. But as things turned out, most of the public never heard about it.

The day after the release was distributed to the media, it seems not to have been reported in a single US newspaper—at least not that I could find any record of. A front-page story in the *New York Times* that day recounted the struggle of a Muslim family to survive in war-torn Sarajevo. But the struggle of all humanity, present and future, to survive on an ecologically torn planet? Nothing. Later, when asked, the *Times* editors said they had considered the story "not newsworthy." The same perplexing response was offered by the *Washington Post*. For reasons that would mystify me even years later, the *World Scientists' Warning to Humanity* had been stonewalled.

Maybe what the *Times* and other papers meant about the document being "not newsworthy" was that it did not contain new research findings; it simply pulled together the broad implications of some very arresting work that had been done over the previous few years but that had been largely ignored by politicians and media. Many of the signers were nuclear physicists, who for years had been maintaining a "Doomsday Clock" showing just how close, in their collective judgment, we had come to the unleashing of nuclear holocaust. Their annual assessments had been treated by reporters as a soft-news item on a par with the predictions of Punxsutawney Phil on Groundhog Day. Many other signers were climate scientists, who two years earlier had issued the first consensus report of the Intergovernmental Panel on Climate Change (IPCC)—the first major warning of approaching climate change. *Their* report had been so heavily disputed by fossil-fuel industry publicists and lobbyists that polls showed the public to be fairly confused and uncertain about whether human-caused global warming—or any global warming at all—was occurring. Others of the signers were concerned about the well-documented history of powerful new technologies being rushed into service and then producing unintended consequences that were too often catastrophic. By the '90s, we'd reached a point where new technology was coming on line far faster than its impacts could be assessed.

Through my work with Worldwatch, I'd become familiar with the phenomena described in the press release, so it wasn't the content that struck me so much as the resolve of so many senior scientists to put their reputations and research funding on the line, even if the *New York Times* and other media would not. The real mystery, in the years that followed, was the general non-interest of the public, even when the issue of human survival was occasionally brought up—as in the rancorous debates about signing an international Climate Treaty (the US Senate voted 95-0 against it). Eventually I just chalked this non-interest up to "denial"—it was something most people just didn't want to think about, as long as the stores were well stocked with food and the gas stations with fuel, and the weather disasters were happening somewhere else.

All that was a quarter-century ago. Now, we're seeing cracks in the denial—invasions of our comfort zones by such events as Superstorm Sandy striking the New York and New Jersey coasts, and the enormous tornado that swept away a good part of Moore, Oklahoma, and the growing frequency of devastating floods in places like Missouri, Iowa, Ohio, and South Dakota. But by now it isn't just climate-related disturbances or other events directly addressed by the *World Scientists' Warning* that have been making life look less stable. Some of the most alarming signs have been related to the digital tech revolution: the thefts of personal data (including Social Security numbers) from at least one hundred million customers of Home Depot, Target, and Anthem Blue Cross; the Kaspersky security firm's belated discovery of a heist from some 200 banks that had smoothly pulled the equivalent of a thousand Bonnie-and-Clydes; and the reports of invisible cyber warfare against the American economy emanating from China or North Korea, even as drone bombings in Pakistan and Syria have emanated from undisclosed locations in Colorado or Florida. Traditional measures of security—the distance from a threat, the sanctity of your own home, the well-guarded bank, the solid reputation of insurance giants, the vaults of the US government itself—all are being breached.

The enormity of our civilization's dependence on tech systems is illustrated by a series of staggering breakdowns that occurred during a single twenty-four hour span in the summer of 2015—breakdowns that officials tried to assure the public were just "coincidence." On the morning of July 8, the United Airlines national computer systems broke down, halting its operations and causing 1,210 of its flights to be either canceled or rescheduled, and disrupting airports across the United States. That same morning, the computers at the New York Stock Exchange crashed—halting trading for hours. Soon after that, the website of the *Wall Street Journal* crashed. A stock-exchange spokesman assured nervous traders that the trading failure was "just an internal technical issue" and "not the result of a cyber breach." The Associated Press reported that the coincidence of these breakdowns "appears to have been a fluke." But a few hours later, the US Government admitted that one of *its* computer systems had just been breached and that hackers had stolen Social Security numbers from 21.5 million Americans.

The assurances of officials that all these events were unconnected rang hollow. And as would later become apparent, such escalating failures are far from being unconnected. All of these systems (and thousands of others) are dependent on the same small group of GPS satellites on which the whole world's communications and electronics depend. And all are part of a pattern that Aviva Litan, an analyst at the information-tech firm Gartner, described as the rude realization that "humans can't keep up with all the technology they have created."

In the quarter-century since the *Warning*, the signs of a deeply troubled world have gotten worse on almost every front, and the evidence of failing life-support systems has proliferated to the point where many of us are losing confidence in the abilities of government or technology to reverse the tide. We're not holding our breaths waiting for the EPA to save the environment, or for the DOD to announce that it is destroying its nuclear arsenal because Russia and China have agreed to do the same. Few of us expect that the islands of plastic trash and rivers of factory-farming runoff forming dead spots in the oceans will get cleaned up, or that cancer or heart disease will soon be cured—or that cancer meds costing over $1,000 per pill will soon be affordable for all. A good many of us are getting more focused now on what we can do for ourselves. With each new alarming sign, there's a growing sense that survival is getting personal.

Since the great institutions of society (governments, multinational corporations, major religions) seem bound (by their ideologies, financial ties, and bureaucratic inertia) *not to acknowledge* that civilization is failing, and *not to address* (except in trivial, don't-rock-the-boat ways) what we can do to get off our collision course, we will have to act for ourselves on an individual, family, or small-community basis. The hard part is that to do that, we need to understand not just what has happened to our modern world but *how* it has happened—what the unacknowledged paradigms of deterioration have been—because now may be our last chance to interrupt those paradigms.

So, my purpose is not to argue that civilization is headed for collapse. I leave that to the scientists, some of the most prominent of whom (physicist Stephen Hawking and biologist Paul Ehrlich, among others) have recently come out and said as much fairly explicitly. In this book, I summarize their arguments briefly. But my larger purpose is to illuminate how we got to this point, not only so that at least some of us can survive but also—more critically—so that those who do may be able to continue the human journey into a less violent and self-destructive future. Along the way, and in later chapters, I raise pragmatic questions we will need to consider in preparing for that epic endeavor. We'll need to dodge the post-apocalyptic cliché of the dazed dystopian survivors thrown back to a "cave-man" existence. It's important that some of us not just survive, but do so with renewed vision, spirit, and strength.

This undertaking should not be conflated with the efforts of the growing numbers of survivalists whose plans seem predicated on the assumption that their main objective when things go bad (or when TSHTF, as they put it) will be to find refuge from anarchy, and to hole up in remote cabins or forts—and who devote little attention to the looming question of where humanity will go from there. What happens when the canned food runs out, and there are no stray cows or deer left to shoot? What happens to the kids?

A key premise of this book is that life is worth living in large part because it is by nature cyclical and ongoing—because many of us have children and grandchildren who are part of a miracle that transcends our own brief moment. If that is so, true survival means something far different than retrogressing to the allegedly short, brutal, and nasty life of a pre-modern or Mad Max existence. Rather, it will mean progressing to a new kind of future that in many ways—we can hope and intend—will someday retain the best of what our evolution and the past 10,000 years of civilization have brought us but will be less crippled by the forces of destruction our current world has loosed on us.

The path to that brighter future has not yet been blazed, so neither I nor anyone else can claim to provide a sure guide. The trek to that hoped-for-life after collapse will be an epic adventure. But by looking closely at the most probable scenarios of dystopia now looming, we can find logical clues to the steps we'll now need to take.

For me, this story is also about what my family, friends, and some colleagues and I—and a community willing to reconnect with the long-dormant skills of human survival, and commit to the long trek ahead—may yet do to thrive. It's ironic that the great global civilization that was made possible by human invention to begin with may now be on its last legs because our thrall with invention has gone too far, too blindly, too fast. But I have reason to hope that for those who wake up from our tech-coddled reverie in time, the human journey may yet go on.

1

Losing Balance

Technologies Tilting the Wrong Way

For the good people of our world, there has always been a certain comfort—and maybe also some complacency—in our knowing that when threats to our life have arisen, they have ultimately been fended off. Our stories—in novels, movies, TV—reassure us: there are bad people among us, and probably always will be, but they get caught or killed in the end. There have been a thousand wars in our history, but we humans are still here—more of us than ever before. The developing technologies and skills that enabled us to build civilization have also enabled us to keep the endless struggle between creative and destructive uses of those technologies and skills in a kind of balance. Only recently—in the last 1 percent of civilization as we can recall it—have we seen disconcerting signs that on a global scale, the balance may be tipping in the wrong direction.

Going into the second decade of the twenty-first century, no one had heard of ISIS. Suddenly, there it was—so shocking to most human sensibilities that countries like Jordan and Egypt, which had remained on the sidelines during the bloody wars in Iraq and Afghanistan, became enraged enough to send bombing missions, and the governments of sixty nations met for an emergency conference. But ISIS was only the latest of those signs, which had begun appearing at an accelerated rate a quarter-century earlier. Until the early 1990s, almost no one had heard of global warming and the onset of "climate change"—a highly euphemistic name for the destruction to come—but suddenly it was upon us, along with the disconcerting message from thousands of scientists that the destruction was being largely human-caused.

Signs of the tipping are omnipresent now, and two measures are unmistakable. First has been the massive build-up of what we Americans were warned about by—ironically—the man who led the Allied victory in World War II, Dwight Eisenhower. The "military-industrial complex" Eisenhower cautioned about is now vastly larger than he likely ever feared it would be, and it has consumed a world of resources that might otherwise have been used for better education, health care, and preparation for the coming catastrophes of climate change and ecological failure. Second, and not just coincidentally, has been the equally heavy build-up of multinational corporations whose industrial waste is polluting our land, fresh water, air, and oceans—and driving our fellow species to extinction—with virtual impunity.

These two overwhelming categories of destructive force far outweigh such shocking phenomena as ISIS, or the hacking of our insurance companies and banks, which are only secondary symptoms of the general shift. Yet, while together they put heavy strain on the ability of civil society to stay in control, they are only the most obvious parts of the shift. There's a third destructive vector that may be the one that finally tips the balance: the unseen, cumulative impacts of hundreds of everyday technologies we use for our convenience, entertainment, and comfort that are in many ways making us too distracted, inattentive, oblivious, sedated, *and weak*—whether as Americans or as citizens of the world—to regain control of our destiny in time.

What We Don't See

As a student at Swarthmore College, a lifetime ago, I studied scientific method—how we know what we know and how we build on that knowledge. That study was fairly academic, and didn't much excite me. Later, in grad school at Columbia, I studied something I happened to have a "real-life" personal interest in—the physiology of the human body in motion. I'd been a competitive runner in high school and college, and the grad work came as a revelation: if science has direct implications for the things you have a passion for, it can become deeply engaging. If not, it drifts to the back of your head—out of sight, out of mind. For me, that was a key insight into the phenomenon of denial, and particularly the enormous denial by most of the American public, of what scientists were trying to tell us. I realized that even something as horrific as the possibility of nuclear holocaust can drift out of sight, out of mind, if it's not related to everyday experience. In the 1950s and early '60s, when I was a teen, and atomic bomb tests were lighting up the sky in US and

Soviet atmospheric tests, we were acutely conscious of the nuclear threat. When the tests ended, public anxieties abated—even though the nuclear arsenals remained poised.

Decades later, while reflecting on some of the more recent research in human physiology, I found myself wondering: if the public fear of nuclear war can recede so far that it doesn't even register in the media lists of top issues, how in the world is anyone likely to get worked up about such small things as weakening muscles due to physical inactivity, or a shrinking hippocampus (the part of the brain that enables mental navigation) due to increasing reliance on GPS? That got me to ruminating about the growing multitude of everyday technologies we now routinely rely on, and on the growing dependence we have on them. I was far from alone in wondering what their cumulative effects might be.

A few years ago, I saw a cover story by Nicholas Carr, in *The Atlantic*, "Is Google Making Us Stupid?" And more recently, I noticed a *New Yorker* cartoon that riffed nicely on Carr's theme: a man and woman are at the altar, and the minister is ready to pronounce them husband and wife—except that, hold on, they're both busy on their phones. And sometimes, reality overtakes comedy: in 2015, the *Huffington Post* published a photo of a man sitting on the deck of his boat off the Southern California coast, looking into his phone. Between the photographer and the man, a Humpback whale is surfacing—so close to the boat that the man could have jumped onto its back. But he doesn't see it! He's on the phone. There was no commentary accompanying the photo, on the implications of such obliviousness for our big-brained species; it was just another amusing moment.

The message of a lot of those digital-tech cartoons—and amusing observations—is that contrary to the popular cliché that tech "connects" us as never before, in significant ways it *dis*connects us—from physical or face-to-face contact not only with each other, but also with the natural world to which we owe our existence. It's not the relatively recent inventions of military weapons or internal-combustion engines or smart phones, but the complex natural world with which we evolved, that enabled humans to survive for hundreds of millennia before civilization began.

The original teacher of survival was biological evolution, so some of the most pointed humor about human survival comes from the somewhat urban-legendary Darwin Awards, given posthumously to people who do such stupid things—who are so disconnected from the skills evolution taught us—that they get killed off before they can reproduce and pass on their "stupid" genes. What we face now is the possibility that our whole civilization may soon be deserving of a Darwin Award—not so much because of defective genes, though, but because of escalating misuse of the intelligent nature most humans are born with.

Having spent so many years immersed in the cauldron of worried science, I see that there's more to this recent amusement about our tech-dominated culture than just easy jokes. A larger reality lurking behind all the smart phone, smart toilet, and Google Glass jokes is that nearly *all* technologies are disconnecting us—not just some of the sillier digital gadgets and apps. The cars or planes we ride in separate us from the earth we ride over. The technologies of food processing separate us from the farms that produce actual vegetables or meat. The technologies of remote military deployment enable an officer at a secure facility in America to kill dozens of people in Afghanistan or Syria without ever having to look into their eyes. The GPS in a hiker's Garmin disengages him from inborn navigational skills that were essential to the survival of humans for thousands of millennia, but that in the hiker may have shrunk from disuse.

Still, you might ask—*so what?* Isn't it a sign of great human advancement that our techs can now relieve us of all the dirty work and drudgery our forebears had to endure? Isn't it a good thing that we can be relieved from hip pain, depression, erectile dysfunction, or a hankering for mindless entertainment or titillation anytime we want? And isn't it great that we can do things so much *faster*—sometimes thousands of times faster—than our parents or grandparents could? Isn't it a blessing that by the early twenty-first century, an Intel chip the size of a fingernail could do in one second a calculation that would have taken one of the early supercomputers *years*? With such awesome advances in power and speed, doesn't our tech development give us vastly more freedom and choice, and more time for the things we really value?

Of course, all those questions are rhetorical traps. Recent studies confirm that Americans today feel *more* pressed for time and *more* frustrated with massive tech malfunctions. Faster *techs* don't always translate to faster work by tech-dependent *systems*. It takes a Los Angeles County deputy just a few minutes to write out a collision report at the scene of a minor traffic accident, but it takes the Sheriff's Department bureaucracy about thirty days to "process" the report. We are also more uneasy than ever about things our techs do that are *not* what we want—Big Brother surveillance, offensive TV commercials invading our personal space, malware in our computers, and scam phone calls undeterred by the federal "Do Not Call" list. Will we soon have camera-equipped drones peering in our bedroom windows, or driverless cars running over our kids?

In trying to understand why such frustrations have become chronic, I have noticed that it is apparently not just our digital devices and diversions, but the products of human invention going all the way back to the First Agricultural Revolution beginning around 10,000 years ago, and then all the technological revolutions that have followed, that are at play here. Through accelerating accumulation, they have led to a state where instead of feeling on top of the world like the champions of survival we humans have been until this century, we now feel vulnerable, stressed out, and apprehensive in ways that past generations probably were not.

One thing I learned early is that when technological change is rapid, the excitement it brings is often tainted by exasperation. As a teenager, I got interested in printing and publishing, and somehow persuaded my parents to buy me an antique cast-iron printing press that was about the size of a refrigerator, weighed a ton, and used the same letterpress method invented by Guttenberg. I learned how to set type by hand, one "a" or "p" at a time, and after setting a few sentences my fingers would be gray with lead dust. Lead was the main constituent of the alloy used to make type in those days, and I wonder how much of it I absorbed. In any case, the process was just too slow, and after a year I abandoned that beautiful machine (I wish I still had it) for much easier methods—first, something called hectograph (making a dozen copies by laying a typed master face-down on a tray of ink-absorbing gelatin); then mimeograph (*hundreds* of copies!), then a field trip to the local newspaper's tractor-sized Linotype machine (*thousands!*).

Years later, I started a magazine about my favorite outdoor activity, *Running Times*, and bought a Compugraphic typesetting machine the size of an upright piano, which actually had *memory*—a whole line!—so if you made a typo while still on that line, you could fix it instantly. Of course, if you went past the "carriage return" before you could stop your eager fingers, you'd have to print out the page, place it on a light table, retype the line that had the typo, print that out, run it through a waxing machine, lay the waxed paper with the corrected line over the typo line, and use a razor blade to overlay the corrected line on the bad one—and be sure you didn't leave it crooked. However, as it turned out, after a couple of years the Compugraphic was totally obsolete and had to be junked, as the wonders of much more advanced word-processing machines beckoned. Then came the siren call of desktop publishing, which for a while overlapped awkwardly with our use of wax machines and light-table layout. Finally, we got the Internet and digital transmission of manuscripts. What had required laborious all-nighters fueled by bad coffee in the pre-Starbucks era could now be done in minutes. Similar progressions of innovation-cum-obsolescence have characterized our progress in sound recording, personal computers, telephones, and a thousand other categories of tech products. Things just keep getting better, or at least faster.

Funny thing, though: I recently talked with the last editor of my old magazine (it merged with *Runners World* in 2016), who confirmed what I have long found puzzling not just about publishing or printing tech, but about technology *in general*: no matter how much faster or more powerful it gets, the use of it doesn't seem to be making our lives any easier, happier, or more fulfilling. If anything, things are more frantic and fraught with malfunction than ever. It's worth asking, now, why we *want* things to be ever faster. From what I knew of human physiology, I began to suspect some years ago that our pursuit of greater speed in all things was proving to be a colossal mistake.

After leaving my magazine and going to work at Worldwatch, with its big-picture studies of global agriculture, water, resource depletion, hubristic World Bank and IMF-funded development, and the emerging studies of biodiversity loss and climate change, I could see commonalities across the fields that the specialists might not. Specialists are by necessity deeply preoccupied with the frontiers of their own research, which gets increasingly narrow as knowledge advances.

Even the personal interest I'd cultivated as a long-distance runner, which I continued to pursue as I grew older, proved to be provocatively interdisciplinary. My early interest in the physiology of endurance piqued my curiosity about our abilities as humans not only to survive but to see ahead on the trail of life—to envision a place we have not yet reached. In the 1980s, I became familiar with the revolutionary "persistence-hunting" theory of human evolution published by biologists David Carrier and Dennis Bramble at the University of Utah and Daniel Lieberman at Harvard—their surprising discovery (earning a cover story in the journal *Nature*) that humans evidently evolved by using endurance and patience, rather than speed, to make their way in the world.

A few years ago, I began a personal investigation of a question that had been puzzling me for years: How does technology, which enabled us to build civilization to begin with, now have us racing toward what many of the world's top scientists fear could destroy that same civilization? What I found may be cause for both Darwin-Award humor and existential alarm:

- We are infatuated with our technologies of effort-saving convenience, entertainment, and sedation, but *the more we indulge in or depend on them, the weaker we become as individual humans.* If I ever stop running, and just ride wilderness trails on a gasoline-powered dirt bike, my legs and lungs will weaken. And that goes for our brains, as well. When we let our devices take over more and more of our thinking and decision-making, our cognition and ability to remember the past and envision the future, too, weaken.
- Because our techs are now so capable of almost anything, we fool ourselves into thinking it is we, the inventors, who have those capabilities. We don't. Very few of us are inventors. In fact, the opposite is true: *The more knowledge society accumulates as a whole, the smaller a share of it any one person has.* In the realm of the Internet, the great libraries of the world, and vast government and corporate databases, each of us is actually more ignorant of our world than the serfs of the Dark Ages were ignorant of theirs. As societal knowledge grows, individual ignorance also grows, and self-reliance shrinks.
- *The more powerful our techs become, the greater the danger that we will misuse them, with catastrophic results.* I first learned this from a very gentle and kindly man who in his younger years had built (for the US military) an atomic bomb that could kill over a million people in less than a minute—but who one day woke up to wonder what on earth had made him do that (more on that in the next chapter). If his bomb had been dropped on Moscow, as tentatively planned, it would probably have triggered a civilization-ending nuclear holocaust. Providentially, that didn't happen. But as my young grandson and a hundred million other kids his age will soon learn, we now have even greater risks of tech-apocalypse than we had then. It's not just atomic bombs, or the weed killer Roundup, or the still rising levels of global-warming gases, now. *It's the stuff we buy to entertain or sedate ourselves, or to relieve us from physical or mental effort, that may be our greatest risk.*
- The *faster* the new techs of the moment are released, the more of our time and energy is squandered trying to keep up, and the less awareness we have of our civilization's epic past and anticipated future. The more we try to "keep up with the Kardashians"—to keep pace with the latest celebrity selfies, smart phone apps, or viral videos—the less capable we are to draw on the wisdom of evolution and history, and to prepare for what's becoming an increasingly perilous future.

The chapters ahead tell the story of how I came to know (or at least suspect) these things, and how my family and I—and I hope you and yours—can survive the dangers they bring. I'm glad to know I'm not alone in my feeling that our thrall with tech is taking too heavy a toll. Just a few years ago, the general public's response to the rise of what we called "high tech" was almost cult-like. Figures like Steve Jobs and Ray Kurzweil were American idols. But then a few prescient critics began questioning where this thrall was leading. In the past few years, we've seen a welling discussion (at least in some of the non-mainstream media) of whether our preoccupation with "tech" is taking over our lives in worrisome ways. Is Silicon Valley playing God? Is Google really making us stupid? Is automation making us incompetent? Is the Internet doing an end-run on the sovereignty of nation-states? Is Big Brother here for real, only more pervasively than even George Orwell imagined? *Are we losing control of our humanity?* But to appreciate that conversation fully, it's essential to see that it's not just about the rise of Amazon and Google and a hundred other startups that have made multi-millionaires out of kids not yet fully cognizant of risk, but about a phenomenon of human invention that began with the agricultural revolution of 5,000 to 10,000 years ago, picked up speed over the past 500 years, and has gone into a hurricane-like acceleration over the past 50 years.

We live in the moment, now. But if you can see beyond the moment, you can also see that the seductions of technology always have a price. In 2015, the Educational Testing Service at Princeton University announced the results of a study that was reported by Andy Campbell in the *Huffington Post,* as follows:

> We American Millennials are really, really good at using today's technology. That's probably because we've forced companies to make technology that's extremely easy to use, and that's better at doing our day-to-day tasks than we are. But with easy power comes no responsibility, it seems. According to a new study, we're some of the least skilled people in the world.

The study, called the Program for the International Assessment of Adult Competence, tested young adults in twenty-three industrialized countries, in three areas of competence: literacy, numeracy (what I call "number sense" in this book), and "problem-solving in a tech-rich environment". The United States ranked twenty-first of the twenty-three countries in numeracy, and eighteenth of twenty in problem-solving.

If our telecommunications networks and power grids are destroyed by a rival country's missiles taking out our GPS satellites, or if the companies and agencies running those networks and grids should fail, Americans stuck with useless devices will be very poor at the kinds of problem-solving needed to survive. The price of increasing tech-dependence is increasing weakness, as a later chapter will detail. And greater weakness means greater vulnerability to being controlled by other people. Heavy tech-dependency is a lot like heavy drug dependency, so in its impacts on society at large, the aggressive marketing of new techs has a lot in common with drug-pushing. It has played a large role in the recent tipping of the historical balance between the destructive and creative uses of technology toward a calamitous outcome.

To those who seek a future of liberation from that outcome, it's not too late to begin the quest. But to go forward with any hope of success, we need to look back at what has caused that unbalancing, so we can avoid making the same mistakes again. The first step, then, is to accept the now very real possibility that it is our inventiveness as humans, the very inventiveness that enabled us to build civilization in the first place, that may now be our greatest vulnerability.

For the Far-Seeing

As humans, we have enormous natural strengths—the products of a hundred thousand generations of evolution. To continue that evolution, we will need to rediscover and practice those strengths, not let them die on the vine as they are now doing.

Those of us who consciously hope for the next few generations to be liberated from heavy tech-dependence, in such a way that technology is used to heighten our experience of life rather than diminish it, can begin that liberation in ways that are below the radar of those institutions that have incentives to keep us mindlessly consuming or sedated.

To begin with, in the course of regular daily activity, find ways to exercise every part of the body and brain—feet, legs, heart, hands, eyes, ears, memory, imagination, navigation, inspiration, calculation. Be aware that most of these skills, *essential* to those 100,000 generations before us, are now largely handed off to our techs. To reclaim these skills doesn't mean rejecting the vastly greater power of a particular technology where greater power or speed is what's really needed, but it means consciously avoiding the unconscious drift into habitual tech-dependence and addiction. The goal is to be highly selective in the use of tech so that as the juggernaut of hyperindustrial civilization gains ever greater speed, we are able to disengage and find alternative ways—more consciously robust, free, and sustainable ways—of moving forward.

Another perspective: when we use powerful techs, they are generally *not ours*; they are owned by their inventors and investors, who profit at our expense, while our own tools of survival go fallow. To prepare for what's coming, let's reclaim our own tools.

2

Big-Picture Thinking

How Some of Us Came to See What We Now See

During the tech tsunami of the past fifty years, initially through sheer happenstance and later through growing curiosity and excitement, I found myself crossing paths with some of the most extraordinary men and women I could ever have dreamed of meeting, and some of whom would eventually be recognized as path-breakers of late twentieth- and early-twenty-first-century human achievement—*and caution.* In revisiting that half-century I compiled a short list of recollections—flashbacks into a life of staggering acceleration in our collective advance—that I thought might give shape or illumination to the genie we have let loose.

Eight Flashbacks

- It was the early 1940s, and the first four years of my life turned out to be the most violent four years in human history, though of course I would not know that until later. I was two months old when Japanese warplanes bombed Pearl Harbor, and I was about four when we repaid Japan by obliterating two of its cities with atomic bombs. I wouldn't remember those four years at all, but would later wonder if the distant reverberations—in Auschwitz, Ardennes, Warsaw, Normandy, Midway, Okinawa, Stalingrad, London, Tokyo, and then Hiroshima and Nagasaki—must somehow have reached across the oceans to affect me in a way I could never shake off.
- In great secrecy, a few years after the war—I was now twelve—in Los Alamos, New Mexico, a young nuclear physicist named Theodore B. (Ted) Taylor designed a bomb—the "Super Oralloy" bomb—that packed thirty-seven times the power of the Hiroshima bomb. It was a feat of nuclear engineering that required brain-boggling math. Neither Bill Gates nor Steve Jobs had yet been born, and the first supercomputers had yet to be delivered; the giant IBM 650 was just being announced. But Ted Taylor was one of those people who could do staggering feats

of math in his head before checking with a slide rule. His calculations at Los Alamos were done with lethal precision. The Super Oralloy (aka Project Ivy King) was assembled and flown to the remote Enewetak Atoll in the Pacific, for a test. The explosion emitted a plume of light that may have become suddenly visible, just last night as I write, from a planet six hundred trillion kilometers away. It was the largest fission atomic bomb ever exploded on Earth. From a distant ship, Taylor and his team watched through dark glasses, exulting. They reported feeling a great rush, just as later scientists and engineers would feel great rushes with the first moon-landing, the first clone of a mammal, and the advent of Steve Wozniak's Apple I. Soon afterward, Taylor's team built another bomb just like the one dropped on Enewetak, all ready to be dropped—if the order came—on Moscow.

Knowing nothing (yet) of this history, in 1969 I went to work for Ted Taylor (who by then had left Los Alamos), editing a journal he was writing about the burgeoning of a fast-spreading nuclear industry whose gangbuster growth was being encouraged by the US government and public alike, but that he now believed to have been a terrible mistake. He'd had a great change of heart about the development of atomic weapons—as had some of the other people who worked at Los Alamos, including the Hiroshima Project's head scientist J. Robert Oppenheimer. It wasn't the technology itself (the powers of which are by nature thrilling) that was haunting Dr. Taylor; it was the potential for cataclysmic accidents or misuse. Taylor had formed a research partnership with my brother Robert (Bob), who was also a physicist. What brought Bob and Ted together, despite their very different backgrounds, was a common interest in the future impacts of technology on our troubled world. Both had experienced existential alarm. Ted, after feeling the euphoria of that almost celestial bloom of light over Enewetak Atoll, had begun to fear that the mad race of the United States and Soviet Union to build nuclear arsenals was becoming an open invitation to terrorists, rogue nations, and black-market crime-lords to wreak havoc on the world. He now regretted what he'd been a part of at Los Alamos, and wanted to devote the rest of his life to helping avert apocalypse. Bob had worked at the Hudson Institute with the notorious Herman Kahn—a futurist known for his devising of nuclear war strategies that would unfortunately kill tens of millions of Americans before the war with the Soviets could be won, and whom Stanley Kubrik said he had in mind when he created the title character for his movie "Dr. Strangelove." Ted and Bob, both feeling that technological development was running amok, began making technology assessments and forecasts for agencies like the National Science Foundation (NSF), International Atomic Energy Agency (IAEA), the Defense Advanced Research Projects Agency (DARPA), and the newly formed Environmental Protection Agency (EPA)—all of which seemed to have been set up as belated public responses to a recognition that tech development was expanding at explosive rates.

- Later that year, while editing Ted Taylor's journal, I also worked on an editorial project with the Wisconsin Senator Gaylord Nelson, chairman of the US Senate Small Business Committee (and later chief counsel of the Wilderness Society), who was concerned about a less ominous but equally world-altering technology: the American automobile. For Nelson, as for Taylor, the danger was not in the technology itself, but in how it was being used. Cars had given us exhilarating personal mobility and feelings of freedom, but were also polluting our air and killing more Americans in crashes each year than would be killed in seven years of the Vietnam War. In 1970, Nelson followed up his auto-industry hearings by joining the environmental analyst Denis Hayes to launch the first Earth Day, which he hoped would provoke a public awakening to the reality that our technologies are having consequences their inventors and investors never imagined.
- In 1991, I moved to Washington to begin my work with Lester Brown at the Worldwatch Institute. I'd been hired initially (before taking on the magazine *World Watch*) to edit the Worldwatch Papers, which tracked research on a wide range of threats to the planet's environmental and economic stability. The first Paper I worked on, by senior researcher Michael Renner, exposed yet another fateful misuse of industry—clear-cutting of old-growth forests with machinery that can rip out in an hour what nature took centuries to produce. Maybe it was then that I began to suspect that a big part of our troubles with powerful technologies is their speed. We don't seem to have much hesitation about getting them launched, but are fairly clueless about how badly they disrupt the biophysical cycles of nature, on which the viability of all life, including ours, depends.

Renner's Paper also alerted me to a phenomenon that would loom large in the years ahead: a great gap between the public's unquestioning regard for technological progress and the warnings from the scientists whose discoveries have enabled that progress. In his Paper, Renner described a bumper sticker that had been appearing on pickup trucks across the Pacific Northwest, where the endangered spotted owl had become a symbol of the endangered natural world at large. KILL AN OWL, SAVE A JOB, the sticker exhorted. In the following quarter-century, I would see variants of that message hundreds of times—as, for example, when a political TV ad appeared in coal-rich West Virginia in 2010, featuring Governor Joe Manchin expressing his contempt for climate science by defiantly firing a rifle into a copy of a climate-protection bill.

By the early '90s, I was seeing an ominous predictability in the American thrall with powerful technologies—whether atomic bombs, internal-combustion cars, or industrial logging saws. When the wisdom of rushing a new technology to market is questioned, people who have an interest in its success are quick to suggest that those who are raising questions are a threat to our economy and liberty. The powers of new technologies can be hugely intoxicating to those who wield them. A common effect is to heighten the confidence of investors—and allied politicians and media—not only in their ability to influence the behavior, beliefs, and desires of society at large, but also in the rightness of their doing so.

- In 1999, as we reached the threshold of a new millennium, *Time* magazine planned a series of essays, "Beyond 2000: Our Planet, Your Health." I was asked to contribute a piece on one of humanity's oldest and most momentous inventions: the domestication of animals. The title was: "Will We Still Eat Meat?" Although I was not a meat eater myself, I didn't take the assignment in order to argue for vegetarianism. Rather, as was my bent, I wanted to turn to the bigger picture—to show how profoundly the advent of livestock has transformed our world. The human population by the end of the twentieth century was a thousand times as large as it had probably been when the domestication of animals began, and it was clear that meat-eating by the majority of that population could not be continued much longer. Industrial livestock production had destroyed much of the world's forest cover (for grazing and growing feed grain), and was hogging (often literally) the world's declining supplies of fresh water. The essay provoked an avalanche of letters to the editor, including one from a top executive of the beef industry, who suggested that the author of this essay was a threat to America.
- In 2010, I coauthored a book with my brother Bob, titled *Crossing the Energy Divide: Moving from Fossil-Fuel Dependence to a Clean Energy Future* (Prentice Hall). One of the most important points we made was that the US economy as a whole was operating at an energy efficiency of about 10 percent (and a gasoline-powered car with solo driver at a *payload* efficiency of about *1* percent!), so there was tremendous opportunity for increasing economic output by increasing industrial efficiency *without* using more fuel. We cited real-world examples of how a few far-seeing businesses were already quietly exploiting that opportunity, and—more to the point—noted that a study by the American Council for an Energy Efficient Economy had found that of all the increase in US economic productivity over the previous thirty years, three-fourths had resulted from increased efficiency, and only one-fourth from new supply. Yet, the country was continuing to squander money and blood on supply-side wars over oil, and lopping the tops off mountains in Governor Manchin's state, for coal, while doing little for efficiency other than encouraging citizens to buy Energy Star appliances and replace their old light bulbs with compact fluorescents.

> The book got strong praise from climate scientists and activists, including Rajendra Pachauri (head of the IPCC, the world organization of global climate scientists), and Bill McKibben (head of the international climate campaign 350.org), but America's consumption of fossil fuels—and emissions of global-warming gases—only continued to rise. Bob and I were discouraged, as I know were Pachauri and McKibben and millions of others around the world. But we also knew that at least some of those millions were beginning to think it is time to look past trying to change the priorities of governments and industries that were too heavily invested in fossil fuels to seriously consider any epochal change of course.

•

Reviewing these flashbacks underscored for me the huge momentum with which the past drives the future. In the current conversation about how Amazon, Facebook, Samsung, Twitter, and the other digital goliaths are dominating our time and lives, one might get the impression that the proliferation of new techs is quite recent, and that the remedy is to pull back from the immersion—get outdoors more, have *one* TV in the house instead of one in every room; cut back on the addictions to Twitter, You Tube, compulsive texting, and checking the email ten times a day.

But as I have learned, heavy tech-dependency is an addiction not just to things like video games or Facebook, but also to coal-fired turbines (which still produce half of our electricity) and gasoline engines, and jet planes and "food" products made by chemists and biotechnicians instead of by farmers. The techs that brought that dependency didn't begin with the whiz kids of Silicon Valley. Before those prodigies even learned to talk, our ancestors' nervous systems were being programmed by the early engineers of nuclear power and biotechnology and refrigeration—and before that, by the titans of the Industrial Revolution, and before that by the ship-builders and navigators of the Age of Discovery. With each revolution, the totality of human knowledge exploded while the share of it held by any one person diminished. And as a result, our *independence*, as individuals, has diminished.

For my generation, with that history in mind, "technology" connoted large, powerful devices or systems designed to hugely increase human capability to build, transport, or destroy. It was about thermonuclear bombs, nuclear power plants, hydroelectric dams, Apollo rockets, big internal-combustion engines, supercomputers (brother Bob and his wife Leslie had both worked on the first two scientific supercomputers), and the industrial robots Bob's future colleagues were developing at Carnegie-Mellon University. I spent seven years with Ted Taylor and Bob, and in that time I came to think of "technology" as whatever we humans have invented to magnify the powers of our bodies or brains. To lift heavy objects, we had built the seven-ton backhoe, the pincer-end of which worked like a giant opposable-thumbed hand. To expand on our ancient ability to throw spears, we had eventually built missiles—which magnified our arm strength thousands-fold and became an "arms" industry. To augment our puny capability with math (not everyone is a Ted Taylor), we built those massive supercomputers. In short, "technology" meant *big*—and that usually meant *physically* big.

Out of Sight, Out of Mind

It wouldn't be until years later that I'd begin to grasp what now seems so obvious: that a lot of the greatest technological advances meant going *small*. The computer revolution would *not* be launched with the advent of the car-sized IBM 650, or the 29,000-pound UNIVAC 1, but with Steve Wozniak's Apple 1—*one-thousandth* the size of those behemoths. The heavy-industry techs of the Apollo moon rockets, General Motors assembly lines, or US Steel blast furnaces would give way to the new fields of microprocessors, neuroscience, and nanotechnology. We were quickly moving into a time when the laptop on your lap was far more powerful than the 14 ton supercomputer had been.

In the thrall of that revolution, few stopped to think that as our techs increased their power even as they got smaller, any potential dangers they carried might also increase. It was an understandable blind spot. Our association of size with power has been built into our cultural memory for millennia, whereas the new inverse relationship—greater power in smaller packages—is still new and counterintuitive. In hunter-gatherer times, people learned that dangers lurked in large animals: tigers, rhinos, anacondas, mammoths, bears. They didn't know about microbes. And even now, that vulnerability persists. In the ebola epidemic of 2014, Liberia's finance minister Amara Konneh lamented that his country was at "war with an enemy we don't see."

In any case, modern humans were not (and apparently still aren't) fully prepared for the idea that tiny techs, along with their huge powers, can also bring huge debilitation:

- We are aware of the voluminous air traffic connecting the world—more than twenty-eight million flights carrying two billion passengers per year, or more than five million per day. We don't see, or think about, how much that massive crisscrossing of borders multiplies the movement of alien microbes in our luggage, bodies or carry-on food, or in the wheel wells of the planes. Yes, we have drugs now that we didn't have to combat the catastrophic flu epidemic of 1918, but we also have vastly more exposure than people had then.

- We have big government agencies like ATF, DEA, FBI, and the border patrol fighting drug cartels, which are themselves equipped with military-grade weapons, vehicles, and communications systems. We're aware of the biological impacts of crack or meth, and of their threats to our communities. But we have little knowledge of the effects of the hundreds of unmonitored and unregulated chemicals or genetically modified organisms we ingest every day, thanks to the proliferating technologies of food engineering, marketing, and advertising. From the standpoint of daily awareness, the products of those technologies are invisible. Who worries about that carcinogenic chemical (4-methylimdazole) used to give Pepsi its brown color, or that other one (azodicarbonamide) used to make shoe rubber and, until recently, used as an ingredient in Subway bread?
- We're very (sometimes obsessively) aware of all the dramas delivered by our entertainment and social media—from helicopters exploding in action movies, to reputations destroyed on Twitter, to the spectacles of the Superbowl or the infectious laughs of late-night hosts on TV. But we have little consciousness of the tiny technologies inside the smart phones at our ears. We know far more about the impacts of linebackers on quarterbacks, or hottie actors on other hotties (all featuring people we don't know personally), than we know about the impacts those devices at our ears (or the content they deliver) may be having on our brains.

The idea that very small, seemingly harmless, technologies can have massively damaging effects may also have been counterintuitive because we long assumed that technology was mainly about doing useful work, and only marginally about things like entertainment and gratification. Half a century ago, after dire revelations about too many workers falling off skyscrapers or bridges under construction, or losing limbs operating factory or farm equipment, much attention was given to workplace safety. There was little concern about people getting hurt while being entertained. But a lot of the time once spent on useful work is now spent on other pursuits.

A large share of our technology is now designed to provide unprecedented levels of convenience, comfort, and new forms of satisfaction, and much less with building sustainable communities, or preparing our kids for the rough road ahead. A producer might now argue that making a video game that titillates the player with fantasies of vampires, assassinations, and exhibitionist sex is useful work because it creates jobs and adds to the country's GDP—which indeed it does. But does it help produce food, water, shelter, mobility, or good health? Does it bring progress in the human journey toward greater understanding of our past, and hopes for the future? And what about the thousands of other gadgets, apps, and distractions that accompany us 24-7? In most, the tech is either very small (fits in your pocket or purse or the dashboard of your car) or invisible (designed to work at a microscopic level). You barely notice or even know of it. But the cumulative effect may be a kind of debilitation of human capacity that has not reached significant levels until very recently.

•

For hundreds of thousands of years before civilization began, the technologies wielded by humans were mainly hand-held weapons and tools—axes, spears, fish hooks, arrows, ropes, baskets, pots. Their purpose was survival. Now our technologies are our nannies. They keep us entertained and satisfied, and they protect us in hundreds of ways—but do little to protect us from the greatest dangers of our age. Their net effect is to distract us from those dangers, and to pacify us. Their overall effect may be to work *against* our survival.

If you're among the richest 1 percent, maybe you can hire a nanny to take care of your baby because the baby will otherwise be helpless when you're not around. *Somebody* needs to deal with that poopy diaper. But of course, as the kid grows older he'll become less dependent and you'll be glad to let the nanny go. A nanny *technology* does the opposite: it increases our power or protection, but only by making us more permanently dependent on its capability to do so—and in doing so eventually makes us weaker, less alert, less able to prepare for what's shaping up to be a very dangerous time to come.

For civilization as a whole, there may be little we can do now to avert calamitous times ahead, as many of the leading scientists discussed in this account now believe. The combination of specialized expertise and generalized ignorance—and the hegemony of industries that profit from that ignorance—impedes any timely reversing of course. But for individuals and small communities who can liberate themselves from the heavy tech-dependence of the power- and speed-escalating culture, there is still hope. That liberation will have to entail something very different than just getting "off the grid," however—and it can't succeed as a generalized "anti-technology" movement. Survival depends on recognizing how tech has *dis*connected us from the world we evolved in, and has tempted us to abdicate responsibility for our strengths and self-reliance.

For the Far-Seeing

To outwit dystopia means not buying into the myth of inevitable tech-driven progress. The gist of that myth is that new techs and the industries that own or sell them will solve any problem that may arise, whether for America or for the world. That "techno-optimism" is a trap, and to get free of it means finding alternative solutions to the routine, reflexive, ways we typically respond to daily needs, *as well* as to unusual crises or challenges.

A quick digression to illustrate the point: When I was editor of *Running Times*, many years ago, we often received letters from people who had been addicted to cigarettes, but who had finally overcome smoking by taking up aerobic running. More recently, in 2015, I watched a powerful series of public-service TV ads depicting a cigarette as a rolled-up contract, which compels the smoker to give up a good part of his or her life in exchange for the smoke. I recalled all those letters we received so long ago, and realized that what those fortunate converts to running had done was to break a life-threatening contract that for too many others had proved fatal.

For those of us who've had too much of our time and energy sucked up by addictions to sitcoms or social media or video games or televised sports, or who've become too reflexively dependent on our omnipresent devices and apps to satisfy every need for pain-relief or assistance . . . there's an analogous opportunity to break the *techno-optimist* contract. In effect, it's an opportunity to replace false optimism with real hope. For a few examples:

- Replace some of the time spent in passive absorption of TV or YouTube entertainment with active participation in a local drama group, or choir, or hiking club, or discussion group. Dependence on media programmed by others is then reduced by greater self-reliance, which is essential to real freedom.
- If you spend long hours at a desk, or riding in a car or plane, consider the recent findings (in multiple studies) that humans often improve the quality of their thinking and problem-solving if they do their thinking while walking or moving around physically. When I worked

with the physicist Ted Taylor, he did much of his serious calculating while walking. Physical activity is not only good for the brain, but good preparation for the day when you might not have the use of all those effort-saving techs you rely on now. Again, self-reliance is a form of freedom.

- Consider that reducing personal reliance on the carbon economy (oil, gas, and coal), is not just an exercise in political correctness, but a means of breaking the very binding contract we have had with the industries that control most of humanity today. Activist organizations like Bill McKibben's 350.org have done a great public service by raising awareness of the fossil-fuel impact on our climate, but the essential next step should be both personal and local:

    - Cut back on driving. If the distance is less than a mile and you're not disabled, walk or ride a bike. If it's raining, wear a hat.
    - Get rid of the gasoline-powered leaf blower, power mower, and chain saw, and for god's sake build a little upper-body strength with a rake, push mower, and bow saw.
    - Better yet, get rid of the lawn, and if you're in an area where water supply is a growing problem, replace the grass with drought-resistant landscaping (xeriscaping).
    - Cut back on foods imported long distances by airplanes that burn gargantuan amounts of jet fuel. In February, you don't need to have summer fruits flown to the United States from South America—and the day will come when you *can't* have them. Prepare for that day.
    - If you have a house with adequate exposure to the sun, install solar power. In my home, even though my solar panels are tied into the grid (to assure power when the sun isn't shining), the installation has cut my electricity bill from around $150 a month to around $5. In the future, I plan to install stand-alone solar, as backup for when the grid breaks down.
    - And so on and on. The carbon economy is like a spider web in which we are trapped like bugs. We need to network with our families and communities, not with the spider.

3

The Numbing

When Our Devices and Distractions Become Our Pacifiers

For the first 99.6 percent of the time we humans have walked the earth (if you start the count about 2.4 million years ago with the first species of the genus Homo), the "environment" has been the physical and biological world we evolved in. For these more than two thousand millennia, successive species of humans were intimately attuned to their surroundings. They had extensive knowledge of their climate, seasons, growth cycles of hundreds of plants and animals, and interdependencies that enabled robust members of most living species, including themselves, to survive. Every boy who survived to father a child, or woman who survived to raise one, had to be highly alert.

If you were an early human, your technologies were little more than minuscule manipulations of the physical world you felt under your feet, and in the breeze or sun on your face and chest. You made hand implements of obsidian, chert, flint, bone, horn, hide, vine, and wood, and after you died, whatever of these you left behind probably took up no more space in the ground than your bones. But around 10,000 years ago, that began to change, and for around 5,000 years the change was fairly slow. Then, in the time after that, the change began to come faster, until in just the past few centuries it became—by the standards of evolutionary time—explosive. In just an eyeblink, we went from the tools of subsistence to the gadgets of power and pleasure—and growing risk of oblivion. For those of us who wish to avoid oblivion, here's a rough overview of the collision course we're on:

- Length of time that humans (including human species that preceded our own) have been on earth: **About 2.4 million years**

- Percentage of our time on earth that humans have been "wild" (prehistoric hunter-gatherers): **99.6**, followed by a 0.2 percent (roughly 5,000 years) transition from Neolithic to civilized.
- Percentage of our time on earth that we've had recorded civilization: **0.2** (since the earliest stone tablets).
- Percentage of our time on earth that has been since the beginnings of the Age of Exploration and then the Industrial Revolution: **0.02** (roughly since the fifteenth century).
- Percentage of our time on earth that has been since the beginnings of the digital communications revolution: **0.002**
- Percentage of our time on earth so far that it now takes for human population to increase by more than it did in the first 99.6 percent: **0.0004** (ten years)

During the past 5,000 years or so (that most recent 0.2 percent of human evolution), our ancestors experienced the first of three successive great waves of technology-enabled disconnection between us and our world—waves of disconnection not unlike what happens if a garden is abandoned to drought. The second and third waves occurred, respectively, in the most recent half-century and quarter-century or so. Here's how it has gone:

Disconnection Phase 1

In the most recent five thousand years, we experienced the first large-scale replacement of nature by technology: reliance on agriculture. The conversion of selected wild animals and plants to domesticated herds and crops meant most people no longer had to be hunter-gatherers trekking through forests or grasslands in search of food or fiber. With the new reliance on a few cultivated species, people became less familiar with the far larger variety of wildlife their ancestors had known. A few farmers could produce enough food for many, enabling the rise of new occupations and increasing division of labor, so most people grew up to be skilled only in their assigned occupations—no longer competent with the comprehensive skills of survival in the wild.

The one-two punch of separation from the wild and separation from previously essential skills meant that along with the greater security of civilization came greater *dependence* on technology. During the previous 99.8 percent of the human journey, technology had served to modestly expand human powers: from throwing a rock to throwing a spear; from cupping water in the hands to carrying a larger quantity in a vessel of wood or clay—and getting a feel for farming.

Then, in the 0.2 percent of our evolution constituting the span of recorded civilization, that expanding of native human powers accelerated hugely: tribes or clans planted larger crops, raised larger herds, excavated blocks of stone to build temples or castles, and devised rock-hurling catapults with which to knock down rival temples or castle walls. Stone tablets *began to record what had been known by forebears*—and the recording of knowledge greatly accelerated the expansions of technology. New inventions not only magnified native human powers (of hands, legs, eyes, and memory) hundreds-fold, but also dramatically altered the physical environment—forests cleared for farming, rivers diverted for irrigation.

The "environment," for growing numbers of people, was now only partly the natural world; large parts of it consisted of roads, bridges, wagons, weapons, tools, granaries, canals, forts, and dwellings. A child growing up in one of the early cities of ancient Macedonia or East Asia would have had far less direct experience of the wild world he or she still depended on, than had his or her prehistoric ancestors.

Disconnection Phase 2

Now jump to the last fifty years or so—the most recent 0.002 percent of our time as exploring and inventive bipeds. What happened to the environment—and to the prevailing experience of it for most people—has been more massively transforming to life on earth than any other phenomenon in the past sixty-five million years. In his 1968 book *The Population Bomb,* Stanford University biologist Paul Ehrlich described what he believed to be a perilous change taking place. Years later, he was widely derided for having predicted massive food shortages and famine in the following few decades, when in fact only relatively isolated famines occurred. But as it turns out, while Ehrlich was wrong about the *timing* (and perhaps too hasty in his predicting the details), his basic message has turned out to be prescient. In the past fifty years, the human population has tripled. It is now growing as much in ten years as it did in a *million* years before the start of the Industrial Revolution. And with that explosion of population has come a commensurate explosion in technology. Ironically, a predominant impact of that *tech* explosion has been to distract us from the ecological impacts of the *population* explosion.

Ehrlich was a key voice in the emerging environmental movement of the 1960s and '70s. But while that was a time when the "environment" became a familiar buzzword, it was also a time when the real physical and biological environment was fading fast from the day-to-day consciousness of most Americans—and with it, any popular awareness of the environmental impacts of exploding population. Even as an alarmed minority of Americans awakened to the damage being done to the natural world, it was the technological environment that was rapidly taking over the consciousness of the majority. Now, as we live our lives in that most recent 0.002 percent of our species' time on earth so far, it's this tech environment, not the natural one, that is most familiar to us, and on which we tend to place our greatest trust to protect us.

It's a *misplaced* trust, because that shift of consciousness has eased us into a second, even greater, degree of separation from the living planet. In ways never known to our ancestors of the first 99.96, technology has enabled us to routinely transcend both distance and time. The barrier of distance has been blown away by our cars and air travel. The confines of time have been broken by TV, radio, and the Internet, enabling us to witness events that happened yesterday or decades ago, anywhere that cameras or recording devices were operating.

A decade from now, people will watch a cop in South Carolina aim his gun at a man who is running away in panic, and shoot him in the back—not a re-enactment, but the actual murder, that moment. Or a precocious teenager might be familiar with the most memorable line of Martin Luther King's "I have a dream" speech, and might also be familiar with that iconic photo of a sailor's kiss in celebration of V-J Day at the end of World War II, but not be sure which came first, or which event was where. With the leapfrogging of time and distance by omnipresent technology, a person's sense of real time and distance blurs. By enabling someone to sit on a couch and become more familiar with the sights of Times Square or the biological hotspots in the Amazon or Azores than with his own local streets or parks, our media overlaid a second phase of disconnection on the first.

This second phase of disconnection gets almost no publicity or public-policy discussion. There are concerned blog posts and talk-show commentaries regarding various *symptoms* of the disconnect, such as the reports that too many American kids can't say whether the Vietnam War came before the Civil War or vice versa, or the observation that many people have little awareness of where their food comes from—or their drinking water, fresh air, or even their electric power. But the causes of time-and distance-disconnection are so omnipresent that we don't see them. More and more, we're letting our cameras, film crews, and browsers do the seeing *for* us.

On a Daily Show segment in 2014, Jason Jones interviewed a group of very hip Google Glass wearers who felt they were being discriminated against for having what one of them described as "basically. . . a cell phone on your face." Shouldn't they be free to go out in public, they asked, without fear of being called names, or refused service in a bar, or even accosted, just because people they encountered thought they were being photographed or videotaped without permission? Jones, with feigned outrage, sympathized: "Since when can't a grown man walk into a child's playground with a secret camera on his face?"

Asked by Jones to explain what the benefits of Google Glass really are, one of the wearers replied that they are "an interface between you and the world."

"Do you hear yourself?" Jones replied. "Those are called *eyes*."

Disconnection Phase 3

The third phase of disconnection, gaining a foothold a decade or two ago and becoming an invisible juggernaut in the past few years, has been so sudden, in evolutionary or even historical time, that it catapults us into unknown, perhaps unknowable, mental territory. It has virtually transmogrified the world we think we know.

For the first few million years of our presence on earth, right up through the mid-twentieth century, most of the technologies we've wielded have had what could be called "real-world" impacts—first, and for many millennia, impacts in real time and close at hand (the distance a child's voice could be heard or a man could run fast with a spear), and later at variable times and often somewhere else (via the second-phase-of-disconnection technologies of transportation and communication), but still physically real enough.

Then, in that most recent 0.002, less than a single lifetime, has come a rush of technologies that transcend not only distance and time but physical reality itself: the virtual creations of TV fiction, movies, videogame worlds, fantasy sports, and even documentary recreations of historic events that may or may not have happened as depicted, but that for millions of people may replace any earlier conceptions.

Back in the heyday of three-network TV, I recall reading an investigative report about women who'd become addicted to daytime soap operas ("As the World Turns," "Days of Our Lives") and for whom the daily melodrama fix had become the most important and real thing in their lives. When a leading character in one of those shows died, one of the addicted women went into deep grieving—apparently having come to believe not only that the story was what we'd now call "reality" but also that the deceased character had been her most intimate friend in the world. Later, that theme was reprised in the movie "Nurse Betty"—and more recently has morphed into a blurring of what constitutes reality in mainstream culture. For how many men do the dramas of televised Red Sox or Lakers games trigger greater day-to-day exultation or dismay than anything happening in their own homes? How did the outcomes of *playing games* become more gripping than the resolutions of real-life conflicts?

What seemed like a bit of a nut case with that grieving soap-opera fan of half a century ago is now so commonplace that it's accepted as a normal part of "life." When the character Matthew Crawley died suddenly in the TV series "Downton Abbey," millions of us were shocked. "I love Matthew Crawley to this day. That's life too," commented a viewer more than a year later.

Actual confusion of virtual reality with physical reality is of course a serious psychological problem for viewers who don't have all their marbles, but the greater problem, affecting the greater number of people, is the routine disconnection. You can suspend disbelief when watching a crime drama or spy thriller and still retain your sanity, yet your brain, nerves, and hormones may also be responding as they would if you were in a forest fighting a predator, or inside your front door fighting a home-invader. In a sense, when you're watching this kind of TV, you *are* facing a home invader. So, even if you're not as daft as Nurse Betty, a big part of your emotional life is artificial. Suspension of disbelief is no longer a temporary vacation from reality, as it might have been for Elizabethans attending a play at the Globe, or for nineteenth-century readers waiting for the latest chapter from Charles Dickens. It's now, for millions, five or six hours of suspension a day.

The danger here is that the all-too-real characters of fictional TV, movies, video games, and the celebrity-studded world of *People* or *Self* often don't abide by the principles of real-world human relationships, or of ecological relationships between humans and other species, or now between humans and the roiling climate that will largely shape our future. Their plot lines largely ignore the principles or probabilities of real-world cause and effect, coincidence, and complexity. Leading characters are far more likely to survive than marginal ones; likeable or good-looking ones more likely than evil or repulsive ones, conscientious ones more likely than sociopaths.

TV shows about cops, detectives, and spy agencies depict a fake world in which murder, treachery, revenge, and psychopathic behavior are vastly more common than they are in real life. (I've also noticed, with some apprehension, that a fairly common type of protagonist in crime shows is the environmental activist depicted as a fanatical eco-terrorist—a characterization that may have been suggested by such unsettling phenomena as the murderous manifesto of the 1980s "Unibomber" Ted Kaczynski, but a vilification of what the great majority of environmentalists really are.)

A lot of these shows portray human relationships (especially among teens and young adults) that are loaded with erotic innuendo, impulsive sex, jealousy, obsession, infidelity, revenge, and comedic romance—but largely devoid of the mundane struggles and satisfactions on which strong relationships are built. (Over the years, I've been pleased to discover occasional exceptions—such as the movie "About Schmidt" or the TV show "Parenthood," which prove that banal real-life dramas can be as engaging as those which have to depict at least one grisly murder or exploding car every few minutes.) An intelligent extraterrestrial trying to understand what kind of world we earthlings live in, by intercepting our TV, would get an absurdly distorted picture. Yet, whether or not any ETs are out there watching, that is very much the picture *we* are getting.

Maybe most damaging of all, in its distortion of our capacity to envision the outcomes of real-life struggles, is the tendency of most fictional stories to achieve "closure"—a satisfying resolution to the story's central conflict or challenge. Far more often than in real life, the bad guy is caught or killed and the good guy survives; the kidnapped child is rescued; the man–woman or father–son crisis is resolved in a satisfying way. The history of storytelling, and now the 24-7 creation of redefined reality, requires climax, resolution, denouement. That requirement is so pervasive that I'm afraid we may subconsciously expect real life to bring that as well. It does not. And that leaves us dangerously vulnerable to what it *does* bring.

It's all too easy to shrug off this fake reality as "just TV." But whether it's TV, movies, or YouTube, that five- or six-hour-a-day virtual world is a bubble, and when you're inside a bubble it's hard to see very far outside. TV has been around for only one human lifetime (the most recent 0.003 percent of our evolution), which has made our experience of the world profoundly different from that of the 35,000 human lifetimes before ours. Is that safe?

We're at a point now where the biggest danger of our media immersion is not that TV or movie fiction is violent or gratuitous, or that video games like "Grand Theft Auto" or "Assassin's Creed" encourage kids to exult in the fantasized act of killing other people or hijacking their cars, but that *the daily experience of millions of people is literally unrealistic*—whether in their vicarious experience of crime via highly selective TV news coverage or in their unwitting confusion of true versus reconceived history. If our understanding of the past is distorted, so is our ability to envision (and plan) a fulfilling and sustainable future. Yet, our nanny-tech media have systemically brought us fictionalized accounts that effectively replace what the best historical or anthropological research might tell us really happened. For example, Steve Coll, dean of the Columbia University School of Journalism, has suggested that our impressions of what happened in the Wikileaks affair—an event with large ramifications for the future of American freedom of information—may be largely shaped, for better or worse, by a movie ("Fifth Estate") whose director's previous work featured two vampire movies. As the British novelist Tim Lott wrote in *The Guardian*, "Virtual space aces physical every time."

As critics have wryly observed, even the "reality" shows are unreal. Fantasy has become the new reality, and millions of people willingly give as much of their time and energy to it as the people of the previous 35,000 lifetimes spent on the hard work of surviving, making a living, educating their kids, and coping with real-life threats or difficulties. Whether it was our prehistoric ancestors tracking and hunting or our great grandparents tilling the soil and digging wells, those previous generations were functioning (though less so as technology expanded) in the real physical and biological environment their bodies and brains were attuned to. The TV/gaming/streaming video watchers are immersed in a predominantly artificial world delivered by an army of digital nannies.

It's wrenching, to me, to see how completely we can be taken prisoner by this home invasion we not only don't resist, but welcome and embrace. I found a poignant example in a letter to the editor of *Road & Track* magazine—a meeting place for people who would be unlikely to think of cars as usurpers of real-life experience. The letter writer, Bernie Kressner, was not unappreciative of the difference between real and artificial experience. He wrote:

> Racing is visceral and primal. Its enjoyment comes from all the sights, sounds, smells, feelings, vibrations, aura and atmosphere. If you have

> to manufacture an artificial *vroom* exhaust sound, then it's just plain phony. If someone watches racing just on TV, then that is their loss. There is no substitute for being there.

I couldn't agree more! But the irony, for me, is that in calling a digital engine noise "phony" and suggesting that the senses of "being there" at a car race are what's real, Kressner unintentionally underscores the extent to which the hijacking of our reality has gone. The passivity of watching an auto race (even there at the track) has replaced the activity of driving a car, which in turn has replaced the long-evolved activity of walking and running. The "sights, sounds, smells, feelings," etc., of internal-combustion engines have replaced the functions of those senses in the world where we spent the first 99.6 percent of our formation. The takeover is virtually complete.

Looking Back: Invasion of the Leaf Blowers

While the nanny-tech conquest happened over many centuries, it reached a critical mass in the quarter-century after World War II, when our defenses were down. Before I explain further, recall the three waves of disconnection we've been through. The American cultural changes that began in the 1950s launched a convergence and acceleration of these waves, in ways that seemed harmless and alluring: a spike in automobile travel ("See the *USA* in your *Chevrolet!*"); a new power mower replacing the push mower for the weary husband recently returned from Iwo Jima; a new dryer replacing the backyard clothes line for the wife. Who remembers what a clothes pin is? These developments intensified the first phase of separation that had begun at least five millennia earlier. Then, soon, the tech aids became more ubiquitous—the automatic garage door opener, the automatic transmission (who remembers stepping on the clutch and shifting gears?), and of course the TV remote (who remembers getting up from the couch to change channels?). Even as our entertainment became more titillating and our business and politics more hyped, our actual participation in the maelstrom of the modern world became increasingly passive. And now, for many tasks or preoccupations, we literally don't have to lift a finger.

For me, what this might mean for our strengths as individual humans took a long time to grasp—maybe in part because we are conditioned by our culture to expect that as we pass our 20s, nearly all our capacities begin slipping into lifelong decline. Our running speed, muscle strength, mental quickness, sex drive, creativity, appetite for adventure—all weaken, long before the more late-in-life failures of brittle bones, chronic illness, and memory loss.

The technological expansions of human powers have been typically heralded by headlines and TV interviews. The weakenings, in contrast, are never announced and rarely discussed, and I initially became aware of them only through what seemed like trivial encounters. One of the first of these encounters occurred outside my home in exurban Virginia, where I lived with my wife and daughter on a forested bend of the Occoquan River, about a forty-five-minute drive south of Washington, DC. It was the kind of leafy enclave that attracted people who value their privacy and independence—and who have the money to protect it. We had a lot less money than most of our neighbors, and felt lucky to be there. Every house was artfully designed and unique, and the yards were aesthetically integrated with the surrounding woods of beech, maple, and oak trees. Many of the houses, including ours, were on the river, which was directly on the route of the Great Atlantic Flyway—a principal migratory route of the bald eagle, the American oystercatcher, piping plover, roseate spoonbill, and about 500 other species of birds. Often, from our back deck, we could see great blue herons on the bank. Our neighbor across the cul-de-sac was an avid bird-watcher who'd counted over one hundred species on his half-acre alone.

I was sitting out on the deck with my morning coffee, enjoying the October foliage, birdsongs, and tranquility, when the peace was broken by the sudden roar of a machine—I thought at first that it was a chain saw, a rare noise in this leafy place where most of the neighbors would never cut a tree unless it was about to fall on their roof. I got up to look, and it was a man a couple of houses away wielding a gasoline-powered leaf blower. He was barely visible in a plume of blue smoke, and I thought of that photo I'd seen of a 1950s family lined up next to a government mosquito-control truck, being happily sprayed with DDT. It was the first leaf-blower I'd ever seen close up, and I watched for several minutes with both irritation and curiosity. I had grown up in the wooded suburbs of northern New Jersey, with occasional family vacations to Vermont, so I knew that getting the leaves off the lawn can be a big job. But we had always done it with a rake. A rake doesn't make loud noise and blue smoke.

Of course, I can imagine what my neighbor might have said if I'd suggested that he use a rake instead of his new machine (I didn't actually have the temerity to do so). "This baby saves a lot of time and effort," he might say—maybe with a bit of forced conviction. In subsequent years, as leaf blowers became more popular, I occasionally found myself stopping for a moment by a public park, or a shopping-center sidewalk, to watch a man wielding a blower, and I felt fairly sure that the thing wasn't really any faster than a rake or a broom. The guy didn't have to do a pulling motion with his arms, but he did have to lug a gasoline engine on his back. And as it turns out, leaf blowers *don't* save time. Men who bought them apparently just assumed they did, because the technology of the blower is more "advanced" than that of a rake. But advanced in what way? The blower belched noxious fumes and carbon dioxide, and enough noise to scare off blue herons and jangle the nerves of any nearby person.

"Leaf blowers are diabolical machines," the medical doctor Andrew Weil would comment years later on his website. "A leafblower running for one hour emits as many hydrocarbons and other pollutants into the atmosphere as a car driven at 55 mph for 110 miles." But its greatest offense may be the bit role it plays as one of a thousand small devices that collectively sap the strength and common sense of their users.

By the time my daughter was grown, leaf blowers had gone from being the newest thing to a gadget every suburban man had in his garage. Sometimes I'd go to Gold's Gym, where there'd be twenty guys doing bench presses and curls. But when I drove or ran through our woodsy neighborhood in the fall, I never saw a man exercising his arms with a rake. The guys would pay good money to exercise at the gym, then would pay more money to *avoid* exercise in their yards. After a while I began to suspect that this bit of irrationality was part of a much larger, less innocent, pattern.

A Forgotten Warning from JFK

Signs of pervasive weakness in the American population—physical, mental, and moral—are not new. They began to appear in the years after World War II, when kids of the post-war generation were given physical fitness tests and alarming numbers of them failed. In a major study conducted over a fifteen-year span, by New York City's Columbia-Presbyterian Hospital, researchers led by Dr. Hans Kraus and Sonja Weber gave physical fitness tests to 4,274 American children and 2,870 children in Austria, Italy, and Switzerland. The results:

- In six tests of muscular strength and flexibility, 57.9 percent of the American kids failed one or more of the tests, while only 8.7 percent of the European kids failed.
- In the strength tests, 35.7 percent of the American kids failed, while 0.5 percent of the European kids did.

Meanwhile, in the next age group up, one of every two young men called up for military service in the late 1950s had been rejected as mentally, morally, or physically unfit. Of course, the Selective Service may have had criteria for moral fitness in those days that most of us would consider misguided today. But overall, the results of the Kraus–Weber study and the Selective Service experience were strong indications that the country might be in some new kind of trouble. In December, 1960, President-Elect John F. Kennedy wrote an article for *Sports Illustrated*, titled "The Soft American," in which—citing the Kraus–Weber and Selective Service findings—he argued that physical fitness is not just a measure of a country's military readiness, but "one of the prime foundations of a vigorous state."

At first glance, Kennedy's reference to a "vigorous state" might have been regarded as the kind of political rhetoric you'd expect from an American president preparing to assume leadership of a nation that was being challenged by a menacing Soviet Union. Young and untested, he was a Luke Skywalker to the Soviets' Galactic Empire, and a bit of bravado was not surprising. Also, for anyone looking back a few years later, "vigor" (famously pronounced "vigah") was a favorite Kennedy word. People easily associated it with the Kennedy clan's love of touch football and other strenuous sports. But while one of his motives for the "Soft American" article was obviously the poor fitness of the military, Kennedy explicitly stated that this was not his main concern. While warning that poor physical fitness in the populace is "a menace to our security," he also noted that "We do not, like the ancient Spartans, wish to train the bodies of our youths merely to make them more effective warriors."

What Kennedy meant by a "vigorous state" was something larger, as he took pains to make clear in the article:

> . . . for physical fitness is not only one of the most important keys to a healthy body, it is the basis of dynamic and creative intellectual activity. The relationship between the soundness of the body and the activities of the mind is subtle and complex. Much is not yet understood. But we do know what the Greeks knew: that intelligence and skill can only function at the peak of their capacity when the body is healthy and strong; that hardy spirits and tough minds usually inhabit sound bodies.
>
> In this sense, physical fitness is the basis of all the activities of our society. And if our bodies grow soft and inactive, if we fail to encourage physical development and prowess, we will undermine our capacity for thought, for work, and for the use of those skills vital to an expanding and complex America.

That was more than half a century ago, and now we can make two very belated updates:

- What Kennedy said about "an expanding and complex America" now applies to the whole world—and to America more than ever.
- The physical conditions of America's kids (and now the kids of Mexico, New Zealand, Canada, Australia, and Chile, among others) are more freighted with obesity, childhood diabetes, and early heart disease than they were when Kennedy raised his alarm.

Today, contemplating the sorry state of America's physical fitness is often an exercise in personal resignation—in itself a symptom of how far we have fallen. Thousands of commentaries have lamented the inertness of our chubby children, not to mention their even more chubby parents, but with no serious efforts to stem the spread. A common response to the trend—at least among those who recognize it at all—seems to be to laugh it off. In a *New Yorker* cartoon, a mother has just spoken to her kids who are sitting on a couch, and one of them responds, "Go out and play? What is this, the sixties?" It's funny, but not a solution. What are you going to do, take the kids' Xboxes away? Anyway, for a lot of kids, even going outdoors won't solve the problem: they'll just sit on the step outside the front door, continuing to gaze into their devices—digitally connected, but in a state of third-phase *dis*connection. According to research by the Kaiser Family Foundation, the average American teen spends over seven hours a day in front of a TV, computer, or cell phone screen.

What's happening with kids is paralleled by what has happened with their parents' work: it has gotten steadily and systemically more sedentary. Half a century ago, there was a lot of concern about automation, the first large-scale displacing of workers by machines. Initially, the people being replaced were mainly factory workers or manual laborers. Today, the replacements are more pervasive, encompassing not just simple skills like lifting or hauling, or assembly-line operations, but more complex skills like medical diagnostics or the remote deployment of military robots and drones. In the work of legal discovery (poring through heaps of documents in the early stage of a lawsuit), automation may now let one lawyer do the work of 500, according to M.I.T. researcher Andrew McAfee, who in 2011 organized a conference called "Race Against the Machine." Even some of the skills required for heart or brain surgery have been handed off to tech assists.

Significantly, automation has not just been a process of technology replacing people against their will. Often, the technology has been *embraced* by those whose skills are being overtaken. A study commissioned by the Federal Aviation Administration (FAA) found in 2013 that commercial airline pilots have become too dependent on automatic systems in the cockpit, and in some cases have developed "automation addiction." Ironically, another study, three years earlier at the University of Essex (UK), found that professional video game players have the "reactions of pilots but bodies of chain smokers." In the FAA case, the nanny-tech involved is a life-and-death safety system, and in the other case it's just a game. But in both cases, the net effect is a weakening of human capacity, whether it be physical fitness or high-level skills.

And now, along with the slacking-off of physical and mental functions that those skills once required, we're having our brains rewired to accommodate all the redefined realities we experience via our media: distorted perceptions of risk, conflict, and resolution. Have TV and movie plots taught us to reflexively expect the risks of street crime to be higher than they really are? When I watch a half hour of news on a network station in Los Angeles (we're almost an hour from the city), it seems that most of the newsworthy events are murders, rapes, road-rage shootings, or robberies, with the occasional overturned 18-wheeler or fire of suspicious origin.

One evening not long ago, I watched the ABC news station in Los Angeles, and recorded the first eight reports. They were:

- A shooting
- Another shooting
- Another shooting
- The capture of an escaped murderer
- Another shooting
- People standing in a long line to apply for jobs
- A police chase ending in a crash

Is this how life really is? Some years ago, I bought a derelict brick townhouse in a marginal neighborhood of Washington, DC—got it cheap because the owner had died and the son who inherited it lived in suburban Maryland and told his real estate agent that frankly, as a white man, he was afraid to come into the city to even look at the property. He was glad to unload it for my ridiculously low offering price. I lived there for several years, went into the alley behind the house every day to take out the trash, spent hundreds of hours crisscrossing the city to work, on foot, day and night—including regular walks or jogs across the notorious fourteenth-Street corridor where the big riots of 1968 had taken place—and in all those years never witnessed a street crime. Once, out of those hundreds of hours, I saw a couple of cops putting handcuffs on a man. I don't pretend that crimes didn't happen, as DC was notorious for its high murder rate. But my nerves had been wired by TV for a very different frequency.

Other testimony about such distorted impressions has been collected by a movement called Free Range Kids, whose members say that the risks of kidnappings and other crimes against kids in public places are grossly exaggerated by news media and politicians. A recent *New Yorker* piece by Lizzie Widdicombe quotes Lenore Skenazy, a leader of the movement, who argues that contrary to public impressions, children are quite safe on public streets in America. "If you actually wanted your child to be kidnapped, how long would you have to keep him outside for him to be abducted by a stranger?" Her answer, based on US crime data: *750,000 years.*

Our brains are not only weakening in their capacity for complex skills like weighing risks, but are being rewired with faulty responses to *real* dangers, even as those dangers grow. George Orwell was prescient in his depiction of a state in which the dangers we fear are distorted or fabricated for corrupt or incomprehensible reasons. Senator McCarthy wanted us to be very afraid of Commies in our midst, even though few of us ever knowingly saw one. And as Orwell foresaw, the identities of the people we're told to fear might change, while the deployment of fear continues in new forms without end. Over the past seventy-five years or so, the focus of our fears has shifted from Nazis to Communists to drug dealers to terrorists, and now increasingly to ideological adversaries in our midst. The more compliant and weak we become, the more easily we can be fear-driven.

A few years ago, these multilayered harbingers of a deeply troubled society—physical softness, mental passivity, and rewiring for a reinvented reality—led me into an exploration of specific signs of societal weakness that I suspected might reveal a larger pattern. As a long-time environmental and science editor, I was most interested in hard empirical evidence, and especially quantifiable evidence. As I tracked the signs, a little like a Paleolithic hunter tracking game, I found scattered bits of information—informational scat—that I eventually sorted into nine main categories of growing societal weakness (next chapter).

And then I was shadowed by a larger question: why so many signs of trouble appearing so suddenly, in just that most recent 0.001 percent of our time so far?

For the Far-Seeing

If it's possible for some of us to live without internal-combustion cars, coal-fueled electric power, or imported fruit *without* being thrown back to cave-man existence (but rather moving forward to a more enlightened kind of life), it's also possible to leapfrog the bulk of the nanny-tech devices and apps that have disconnected us from the physical world. If the first step in our journey is to recognize the eventual costs of new technologies that almost always have unforeseen (and often catastrophic) consequences, the second step is to recognize that in the enormously accelerated tech development of just the most recent half-century, those costs have spiked. The weakening impacts of excessive tech dependence, especially among Americans, have become epidemic.

To leapfrog the current insanity, though, does not mean jumping over or end-running the current time in human history. Far from becoming disengaged, it means becoming more actively and consciously involved in the life of our world than we ever have before. It means reconnecting with the natural world we evolved in, both as an antidote to the disconnecting effects of second- and third-phase technologies and as a form of training for the coming time when many of those techs will fail.

Practical suggestions:

- Do more of our living and working outdoors, to become more knowledgeable and at home with the natural light and life of the places where we live.
- Increase physical activity beyond just walking a mile instead of riding in the car—whether by hiking, running, mountain climbing, or working on farm, garden, construction, or renovation projects with hand tools.
- Limit use of digital devices to a very brief period each day, with the expectation that the time will come when your contacts with other humans will have to be mostly face-to-face.

4

The Weakening
Unnoticed Effects of Effort-Saving Technology

If JFK's ghost is out there somewhere, I wonder what he must think of his country now. He might find a particular irony, and heightened meaning, in his famous admonishment "*Ask not* what your country can do for you. . . ." Since his death, Americans have done quite the opposite: We *have* asked—and asked, and asked, and gotten, and gotten. We haven't gotten a lot of what we actually need (good education, good health, and a more stable environment), but we've gotten a Caesar's feast of consumer assists and satisfactions. I find myself cringing as I write this, because I'm the farthest thing from an anti-government, anti-social services fanatic. But I can't help recognizing that we now have what some are calling a "savior state."

Roughly speaking, as of now, only about a third of all Americans over eighteen are actively doing serious full-time work to make a living. The other two-thirds are a mix of still-dependent youths, retirees, and people who are disabled, unemployed, half-employed, hospitalized, institutionalized, or incarcerated—as well as the many who were previously counted as unemployed but have stopped looking for jobs. The nonworkers also include a fair number of the very rich—people who can essentially sit back behind their security gates and watch their investments pay their way, at least for the moment. Gallup's annual "payroll-to-population" figures tell us the nonworking population of the United States is about 57 percent, but that doesn't include the many millions who for whatever reasons are no longer counted as being in the workforce. It also doesn't include those who are on payrolls but don't actually do any useful work because they are dead-wood employees of corrupt bureaucracies. So, the actual percentage is much higher than that fifty-seven.

Some of the nonworkers are among the roughly forty million Americans who will have experienced poverty by the time they're sixty years old, while others are somehow driving around in BMWs. In an article titled "Rise of the Nonworking Rich," former Secretary of Labor Robert Reich noted that in the United States, income from inherited wealth is accumulating faster than income from work. Whatever the numbers, it's safe to say a large majority of our people, including millions who are (at least for now) quite comfortable, are dependent on the work of a minority to survive.

This is not a rap against all the people in America who don't or can't do productive work, however—many of whom would desperately like to have a paying job. It's only a dispassionate observation that conditions are vastly different now than they were for the first 99.6 percent of human evolution, during which most nonworking people would not have lived long enough to pass on their genes. I feel sympathy for people in many of those dependent categories. Some of my own family members, the people I care about most, are among the dependent 240 million or so. And I have *empathy* for *all* of those 240 million, even the freeloaders and crooks. That doesn't mean I'd protect them the way our country protects them, though. By empathy, I mean that I can understand on a visceral level why some of the worst people do what they do. I have good reason to believe every human alive is capable of murder, elder abuse, child abuse, robbery, treachery, terrorism, or just about any other wrongdoing you can name. We humans carry predatory genes, and while culture subdues or sublimates their overt expression in most of us, certain conditions or combinations of conditions can trigger them. When that happens, news reporters quote a relative or neighbor saying, "This isn't the George I know! He just snapped!" Today, we're not so far from having our whole country snap.

This huge dependence of perhaps two-thirds of the population raises a moral dilemma of unprecedented magnitude—and it's a dilemma whether you're an anti-welfare conservative or someone like me who'd be labeled something else, although any label you give me will be wrong. The dilemma is that (1) one-third of the population can't carry the other two-thirds much longer at current levels of consumption without bankrupting the country and causing civil collapse, *and* (2) the dependent two-thirds can't just be abandoned (or even cut back much further than they've already been) without gutting consumption and the GDP, and causing civil collapse. Either way, the status quo is an unstable system ready to collapse. Humanity (not just America, now, but the world) is headed for a colossal train wreck unless we make an epic change of course—a change that gets vastly more people working creatively and hard for our collective survival. Scientists aren't just talking about destabilized climate and rising rates of biological extinction, now, but about the impacts of those epochal changes on a weakened human race.

If there were no alternative to an impending collapse, I might not expend energy writing this; I might be like one of those "Doomsday Castle" survivalists, preparing for the end—though god knows what they're living for, if everyone outside their walls is an enemy and the future is Hell. If there's a real solution for us, it can't be retreat, like that of a prepper in his mountain cabin with a hundred cans of pork and beans and a shotgun; it has to engage us all, at least until our global numbers decline by two or three billion through natural attrition—birthrates finally falling below natural death rates, and those who survive getting a lot smarter than *we* have ever been. That could take several generations.

The solution can't be summarized in a two-sentence Tweet or thirty-second public-interest TV spot, nor is it "simple" in the sense that lawmakers want their signature achievements to be simple. Movie plots and Hallmark card sentiments are simple. Real life is maddeningly complex. One way to grasp what a real-world solution to the dependence dilemma must be is to review, in turn, each of the major signs of growing societal weakness I've been tracking. My expectation is that for each of these signs, one of the keys to remedying the weakness will be finding a way to liberate ourselves from a particular form of technological overdependence or abuse.

Nine Signs of a Weakening

Sign 1: Unfit Bodies

In his "Soft American" article, John F. Kennedy gave pointed emphasis to the connection he'd seen between the decline of physical fitness in Americans and the growing threat to the citizens' *mental and moral* fitness at a time when clear thinking was essential to survival. In an almost quixotic hopefulness, he wrote, "this growing decline is a matter of important concern to thoughtful Americans."

In the more than half-century since then, the decline has continued, and it now appears that Kennedy may have overestimated how many of us are thoughtful. A study by the Harvard School of Public Health reports that physical activity by Americans—at home, work, and school—has declined sharply in the past half-century (that most recent 0.002 percent of our time so far on Earth). During one brief span of that time, between 1969 and 2001, the share of American kids who walked or rode their bikes to school dropped from 40 percent to 13 percent. And that was before the obesity epidemic really hit (see sign 2).

Another study, from the University of North Carolina at Chapel Hill, found similar declines in physical activity in China, India, Brazil, and the United Kingdom. Those countries, along with the United States, account for nearly half of the world's population, so it seems clear that the decline in physical activity—and consequently in fitness—is going global. As nations have industrialized, they have increasingly freed their people from physically demanding work—but have commensurately made their people more tech-dependent.

The UNC study's director, Barry Popkin, noted that in recent decades the introduction of home technology, including rice cookers, stoves, refrigerators, washing machines, and microwaves, has reduced the time traditionally spent producing food and doing housework. That's the good part. But those advances have also brought a reduction in physical activity. The United States has led the world in such advances—and in such reduction of physical activity. I'm thinking about my home state of California, which has been celebrated for its presumed embrace of outdoor living and sports. In 2011, a basic fitness test was given to the state's fifth, seventh, and ninth graders, and 70 percent of the kids failed.

I know this next observation will make me sound like I'm from another century (well, I am!), because when I went to college no student had a personal computer or TV in his or her dorm room. Now, as I understand it, a typical student will have not only a laptop but a tablet, smart phone, recharger, mini refrigerator, coffeemaker, microwave, music system, speakers, earbuds, PlayStation, *and* TV. What else? Maybe a Garmin or Fitbit, or for some of the more weight-challenged, a glucose monitor. I wonder, how much time does the student spend in an electronic trance, which half a century ago might have been spent playing Frisbee or touch football, or walking into town with friends, not to mention spent on serious reading and study?

To put that kind of question in perspective, the UNC researchers developed a measure called "metabolic equivalent of task" (MET), which records the amount of energy expended on a task. The study projected that if present effort-relieving trends continue, by 2020 the average American adult will spend 190 MET hours per week. The researchers then compared two other key estimates: the energy expenditure of a moderately active adult who works at a desk job and does vigorous exercise thirty minutes per day, and the base metabolism of someone sleeping for twenty-four hours a day, every day. Here's how the numbers compared:

Weekly Metabolic Equivalent

- Moderately active adult: 240–265 METs
- Full-time sleeper: 151 METs
- Average American in 2020: 190 METs

In other words, the average American will soon be *metabolically closer to a person in a coma* than to a person we'd now consider just moderately active.

Noir joking aside, the problem with poor physical fitness is not just about its impacts on health and longevity; it's also about the fact—asserted by JFK but now confirmed by hard evidence—that poor physical fitness means poor mental fitness too.

One of the most compelling sources of that evidence was a study of 1.2 million Swedish boys over a three-decade span, in which those who had improved their cardiovascular fitness between ages fifteen and eighteen (by doing cross-country running or skiing) were found, in later years, to have higher IQs, levels of education, and income than those who had not had such exercise and whose cardiovascular fitness had declined.

The study, led by Maria Aberg of the University of Gothenberg in Sweden and Nancy Pederson of the University of Southern California, was published in the *Proceedings of the National Academy of Science* in 2009. The authors noted that "positive associations with intelligence scores were restricted to cardiovascular fitness, not muscular strength, supporting the notion that aerobic exercise improved cognition through the circulatory system influencing brain plasticity." In every measure of cognitive functioning the researchers analyzed—from verbal skills to geometric perception to mechanical skills—average test scores increased according to aerobic fitness.

It's not a stretch, then, to suggest that too much dependence on the car rather than the bicycle and legs, and too much time spent with digital experience in preference to outdoor activity, will have the long-term effect of eroding survival skills and intelligence.

Sign 2: Obesity

The US National Center for Health Statistics reported in 2013 that by 2008–09, five times as many kids were obese as in 1973–74. Since then, the obesity epidemic has continued to spread. As a sign of societal weakness, obesity is closely correlated with declining physical fitness and blood flow to the brain, but brings its own heavy burdens as well. Among Americans age twenty and older, at least *half* are now overweight or obese (we're still at "or"). According to a 2013 report by the Trust for America's Health and the Robert Wood Johnson Foundation, if current trends continue, half of the adult population will be fully obese by 2020.

Tech connections:

- The car, carrying kids to and from school (see sign 1 above).
- The texting and social media, often taking over from actually walking or bicycling someplace to meet with friends or play sports.
- The TV (or YouTube), bringing commercials for Coca Cola, Hershey Kisses, Magnum chocolate, Pepsi, Dairy Queen, Dove chocolate, M&Ms, Snickers bars, and other carriers of concentrated corn syrup or sugar, while also bringing easy entertainment that doesn't require getting off your rounded behind.

There are a lot of other such connections (food engineering for shelf life and mouth feel, or package design for maximum "pick-me" impact, etc.), but these big three—cars, smart phone, TV—have two big impacts in common: they make it easy not to walk, and they make it easy to replace the kinds of food we ate for the first 99.8 percent of our evolution, and to which our bodies are adapted, with highly alien substances that are addictive and cause quick accumulation of fat. The TV (or computer, or tablet) keeps delivering those junk-food ads; the smart phone apps make it easy for the kids to decide which junk-food place to meet at, or where the nearest one is; and the GPS-guided car takes them there. Even kids who do sports seem to ride everywhere in cars when not actually on the field or trail. In 2013, I saw a photo of the players on a top-ranked US major-college football team standing in a row for the camera, and two-thirds of them had bulging bellies. And those were some of the presumably *fittest* young Americans.

A while back, I read a news story about a man believed to be the oldest man ever known to have walked the earth (at least as determined by official birth records): a leathery indigenous Bolivian named Carmelo Flores Laura, who was reportedly born in 1890 and was still working as a sheep herder in 2013, walking up and down mountain pastures all day at the age of 123. He died in 2014. Asked that obvious question all centenarians seem to be asked ("what is the secret of your long life?"), he replied simply, "I walk a lot. I go out with the animals." As far as I could determine from the truncated news stories, Carmelo Flores Laura didn't have a car, smart phone, or TV. There may not have been a McDonalds or Carl's Jr. within a hundred miles of his home. Needless to say, he wasn't fat.

Sign 3: Shrinking Attention Spans

This is one of those aspects of modern life, along with the rush and the pressure, that educators worry about but do little to change. A teacher trying to maintain order in a classroom might be an ad-hoc exception. But society at large? If the stockpiling of nuclear bombs or destabilizing of the climate doesn't particularly alarm most people now, why should such a trivial thing as shortened attention spans? One answer, ironically, might be that distracted attention has reduced people's ability to pay enough attention to those larger issues to know what's at stake.

The more far-reaching problem, beyond chaotic classrooms, is that as the challenges to our survival grow larger and more complex, the time it takes to understand and deal with them safely gets longer. When a critical task (landing a crippled airliner, or performing a risky heart surgery, or negotiating with a terrorist holding hostages) requires staying on task without faltering for many minutes or even hours, it's not a good thing if the person performing the task has trouble staying focused. The need for focus in the face of a destabilizing climate and endangered human species is infinitely greater.

Yet, the tech environment we have now is not conducive to prolonged mental focus—and in fact is just the opposite. If you spend more of your time texting or tweeting or on Facebook or Vine, your brain is wired for shorter, more simplistic messages than those used by the previous generation—and you'll find yourself getting antsy and impatient with longer forms. Unfortunately, that also means you'll be disinclined to read or think through complex material that can't be reduced to sound bites or Tweets. Oh well, WTF.

Advertisers, politicians, preachers, publicists, political "think-tank" consultants, and others who make their living by influencing people's beliefs or behavior are well aware of the rewiring, and adapt to it by making their messages even shorter. Here are some of the changes:

- The average length of a TV news excerpt of a politician's speech had shrunk from forty-three seconds in the 1970s to nine seconds by the turn of the new century, and since then has been reduced to a single sentence or sound bite.

- The average time a person watches a YouTube video shrank from seven minutes in 2010 to five minutes in 2012 and is probably more like three minutes now.
- TV ads have shrunk from mainly sixty seconds a few years ago to more spots of thirty seconds in recent years—and now even shorter. Half of all packaged-food and fast-food ads are now just fifteen seconds. On some TV shows, a single commercial break will bring a blitz of nine or ten fast-action ads with no pauses between them.
- On the popular website Vine, looping videos are limited to six seconds—long enough, evidently. In 2013, a teenager named Nash Grier began posting peeks at his not-especially-unusual teenage life, and attracted nine million followers.

The symptoms of one condition can become causes of another. Rapid-fire communications are symptoms of shortening attention spans, but also are causes of declining ability to watch those probing half-hour discussions I remember with Bill Moyers, or the probing reporting of Amy Goodman on Democracy Now!, or Rachel Maddow on MSNBC—or even those fairly quick-comment PBS conversations with Mark Shields and David Brooks. We now have a huge body of evidence that Americans are not inclined to spend much time thinking about the most critical threats we face, because those threats are frustratingly complex and making sense of them demands sustained reading, thought, discussion with others, and social action. In a discussion of how modern technology rewires the brain, the director of Stanford University's Impulse Control Clinic, Dr. Elias Aboujaoude, warned:

> The more we become used to just sound bites and tweets, the less patient we will be with more complex, more meaningful information. And I think we might lose the ability to analyze things with any depth and nuance. As with any skill, if you don't use it, you lose it.

When it comes to complex issues, the American mind has weakened. It may be little comfort to know that increasingly, so have the minds of other peoples around the globe.

Sign 4: Declining Self-Reliance

Self-reliance was an essential part of the American identity at least up through the early 1940s. The lore of America's role in World War II is replete with stories of men overcoming adversity by drawing on their ingenuity, mechanical and electrical skills, and *can-do* attitudes in situations where the normal protections (fortifications, equipment, vehicles, weapons) were not enough.

Then, in the 1950s, there was a kind of relapse—maybe a combination of letting go after the tensions of war and shifting spending from the war to the new suburban life—from weapons to washing machines. The '50s were such a relief and so welcome to a war-weary public that the initial erosion of self-reliance was not really noticed. And no alarms were raised, because the benefits of convenience and ease were conspicuous (soon leading to conspicuous consumption and "keeping up with the Joneses"), so the losses were at first invisible. It wouldn't be until after the election of JFK that one of the first major alarms—the "Soft American" article—was sounded.

Since then, though, the same core groups of techs that have driven the decline of physical fitness and the burgeoning of obesity—cars, TV, and digital devices—have allowed the atrophy of self-reliance. The atrophy has been further facilitated by a growing dependence on government to keep us safe. With cars came a proliferation of traffic regulations, red lights, auto insurance, registrations, licenses, smog tests, speed traps, radar guns, seat belts, air bags, blind-spot protections, GPS navigation, automatic parallel parking, anti-skid braking, and the early stages of self-driving cars. And along with all of that, we got the bureaucracies of the National Highway Traffic Safety Administration, state highway departments, DMVs, traffic police, collision reports, and municipal courts.

Imagine a John Wayne-style cowboy being stopped by the sheriff for riding his horse past a rural stop sign without stopping. A lot of Americans, even now, chafe at what feels like too much government regulation. But that feeling, now, is more ideological sentiment than reality. You rely on the red light to keep you from being T-boned; you no longer rely on the peripheral vision that was critical to your hunter-gatherer forebears. You rely on the no-wrinkle fabric someone developed in the 1950s to look sharp when you get dressed in the morning; you don't rely on yourself to iron your pants. You rely on the thermostat and central heating to warm your house in the morning; you don't rely on yourself to fire up the wood stove. There have been hundreds of these small conveniences—each relieving us of a bit more of our need to be independent. According to the FDA, fifty million Americans rely so heavily on pain pills, for every imaginable discomfort, that large numbers of patients have become drug abusers. Self-reliance is mostly a myth.

Sign 5: Declining Grasp of Math and Science

You've probably heard some of the numbers. From a recent "Trends in International Mathematics and Science" study, which tested students in more than a thousand schools across the United States, and comparable numbers in other countries, here's a sampling:

- *Rank of country's kids in 4th grade math*
    - South Korea: 1st
    - United States: 11th
- *Rank of country's kids in 8th grade science*
    - Singapore: 1st
    - United States: 10th
- *Percent of students who reached the advanced level in 8th grade math*
    - South Korea: 47 percent
    - United States: 7 percent

And for young American adults, the record is even worse. From the International Assessment of Adult Competence reported in 2015 (Chapter 1), ranking the competence of young adults (born after 1980) in twenty-three countries, in "numeracy" (the ability to understand and work with numbers)

- *Rank of country*
    1. Japan
    2. Finland
    3. Belgium
    21. United States

The trouble with this picture is that if we really value our technology (and obviously we don't just value it but virtually revere it), we'd best not forget that good technology depends on good engineering, good engineering depends on science, and the ability to make good use of science depends heavily on math. Without strong capability in math and science, our engineering will be haphazard and our technologies will fail us more and more.

Recall the haunting list of epic tech disasters: thalidomide, lead-based paint, asbestos, DDT, Bhopal, Chernobyl, Tegaserod, Love Canal, Fukushima, chlorofluorocarbons (CFCs), and hundreds more. Every one of those terrible outcomes could probably have been prevented by more competent applications of the math—whether in the underpinning engineering or (more typically) in the assessment of probable environmental and health impacts and in the resulting political discourse. Too many of our legislators, governors, and mayors, but also too many of our tech people (especially those who got their training at for-profit tech "universities"), were among those who did not do well in eighth-grade math.

Since many of the best schools and colleges in the world are in the United States, an obvious question is why our kids' performance has been so weak. There's been endless debate and tearing of hair about that, mostly focused on our teaching methods and school districts' curricula, and on budgetary tugs-of-war. Little attention is paid to the addictive popular culture and rushed economy in which these schools try to function. If kids are now wired for short attention spans, short cuts, and tech support for almost everything they do, the sustained focus and hard work essential to learning math and science is alien to their rewired nature. There are bright kids living in Cambridge, Massachusetts or Palo Alto, California—the Meccas of math, science, and digital engineering—who get Cs and Ds in eighth-grade math and science. Those towns also have abundant services for good living: Whole Foods Markets, first-rate hospitals, art museums, bike paths, nice parks. Kids in these towns can openly read books or study for tests without getting beaten up by pubescent thugs. So, why are some of the kids in these towns math and science doofuses?

I found one painfully obvious answer when I read an article in the Princeton University humor magazine *The Princeton Tiger*. Here's the title that caught my eyes:

"Decline in Sciences and Engineering Majors
Due to Math Being Hard"

The article described a study which had found strong evidence of a "causal relationship between math being hard and people not wanting to do it." My reaction was: That's no joke! Then I stopped myself and mused, has it come to this? Students choose their majors—and goals in life—according to what's easiest?

But I also thought: why, then, do at least a few million people in this country choose to do things like run a marathon? Running 26.2 miles may not necessarily be "grueling," as reporters who'd never done it often wrote some years ago, but I can attest that it's not easy. I recall reading about a man named Glenn Cunningham, whose legs were badly burned when he was eight, in a gasoline explosion that killed his older brother. His doctors recommended amputating both legs, but providentially his parents refused to approve the operation. The aftermath was arduous; after taking two years to regain his ability to walk, the young Cunningham began to *run*—and eventually made the US Olympic team. In 1936, he won the silver medal in the 1,500 meters at the Berlin Olympics, where he roomed with the legendary sprinter

Jesse Owens. Cunningham didn't become a great runner because it was *easy*, but because for a kid who'd been told he'd never walk again, it was so heroically and thrillingly hard. And Cunningham was far from being an American anomaly. President Kennedy, a quarter-century later, said, "We choose to go to the moon [and take on the challenges of space flight] not because they are easy but *because they are hard*, because that goal will serve to measure and organize the best of our energies and skills." (italics added)

But if the *Princeton Tiger* article is right, a lot of Americans are no longer motivated to do things that are hard, as so many of their forebears did in crossing the Great Plains by covered wagon, homesteading with nothing but hand tools and horses in the Bitterroot Mountains of Montana, or, for many who lived later, working hard for thirty years to build savings and put their kids through college. For most of us, those feats of lifetime perseverance are things of the past. We want quick rewards for minimal effort now (Chapter 6), and have little sense of how unlikely those dangled rewards actually are, in part because we never learned much about *probability*. It's not just companies with get-amazingly-strong powder or get-rid-of-wrinkles cream to sell, but every state government that has approved a lottery, or legalized gambling, that is cynically exploiting its residents' math weakness. We've gotten far weaker in math than a nation with as much tech power at its disposal can afford to be.

Sign 6: Rising Incidence of Mental Illness

A century ago, mental disabilities were considered a weakness in the human race. A lot of the "insane" were sterilized so they wouldn't reproduce. Now we know better—mental disabilities are illnesses, and we try to treat them, although the reality is that ever since a lot of that treatment came to an end as a result of new policies in the 1970s and '80s, legions of the afflicted have simply moved from record-keeping institutions to the streets. Now, instead of having lunatics locked away in places like New Jersey's Greystone Park (a prison-like insane asylum where I worked as a white-suited "orderly" for one summer when I was nineteen), the madness has gone underground—or under tarps, newspapers, and cardboard boxes, when the hour is too late for begging or mumbling in public.

But while mental illness is now officially not a weakness, *the conditions that have allowed the incidence of such illness to rise* in the United States are a growing weakness. A few stats, counting the people who are still known to health officials and have not disappeared into either the purgatory of Skid Row or the shaky camouflage of "normal" American life, where too often we hear that breaking news about someone who "snapped":

- An estimated forty million American adults have been diagnosed with **anxiety disorder**, another seven million with **generalized anxiety disorder**, and six million with **panic disorder.** That's not counting anxiety-afflicted kids under eighteen, whose numbers are exploding.
- About six million American adults have been diagnosed with **major depression**, over two million with **dysthymic disorder** (chronic mild depression), and about three million with **bipolar disorder**—the one we often hear mentioned in those reports about someone who snapped. Two members of my family are bipolar, and I can attest that it's a terribly debilitating—and sometimes dangerous—affliction.
- Between two million and three million American adults are **schizophrenic**—more than one of every hundred of us. A couple of days after a contractor with Top Secret clearance snapped and murdered thirteen people at the Washington, DC Navy Yard in September, 2013, investigators belatedly discovered that he'd been hearing voices and exhibiting clear signs of schizophrenia. He had never received treatment.
- About fifteen million adult Americans are afflicted with **social phobia**, two million with **agoraphobia** (fear of public places), and nineteen million with **specific phobia** (such as abnormal fear of heights or of flying).
- At least eight million adult Americans suffer from **post-traumatic stress disorder (PTSD)**, and the number is rising as belated diagnoses slowly roll back fears of social ostracism or ruined careers among Iraq and Afghanistan War vets.
- According to the National Institutes of Mental Health, about 4.4 percent of the American population—about fourteen million people—will suffer from **anorexia**, **bulimia**, or **binge eating disorder** at some stage of their lives.
- About three million adult Americans suffer from **antisocial personality disorder**, five million from **avoidant personality disorder**, and another two million from **borderline personality disorder**. That last one is a doozie—contrary to its name, a witch's brew of multiple afflictions that can turn a seemingly normal person into a chronic monster.

These numbers just show a moment in time—the latest assessment at the time I looked them up. By the time you read this, they may be larger. But don't add them up, lest you mistakenly shock yourself into thinking fully half the US adult population is mentally ill. It's not that bad. Some people have more than one diagnosis, so these numbers I've listed show considerable overlap. A person very close to me, for example, has been diagnosed with bipolar illness *and* anxiety disorder. Yet, this individual has admirable bursts of sanity and social strength and I believe will be an asset to society. On the other hand, with its usual assiduous analysis of the data, the National Center for Health Statistics has determined that all told, about one of every four American adults is mentally ill. And, to repeat—and as the experts who compile these statistics caution—the official numbers may grossly underestimate the real depth of the problem. Millions of the people among us are disturbed but remain under the epidemiological radar. The real number may be more like one of every three.

Bottom line: we agree that mental illness is not a weakness, but whatever is causing so much of it in America clearly is.

Sign 7: Rising Incidence of Suicide

A suicide signals that someone has rejected the instinct to survive that has driven evolution—an instinct that I'd like to think remains intact in most of us. If rates of suicide are rising (and they are), it's worth asking whether that primordial instinct has weakened in society as a whole. Does the *visibly* weakening grip on life by an iceberg's tip of humanity signify some larger underlying growth of ennui or fatigue, or deepening pessimism about the future?

Or, perhaps, does it partly reflect an unanticipated effect of embracing technologies that make it much easier for people to do things quickly, including things like stopping pain, escaping fear, or bringing closure to unresolved conflicts? There's a plethora of nanny-techs specifically designed (or at least often used) for attempts at quick relief or release: Oxycodone; Aleve, Budweiser, Jack Daniels, Camels, ear buds, meth, and red Corvettes. And beyond those, we have a growing arsenal of tech-supported actions aimed at sudden transformation—from fat to thin, unknown to famous, financially distressed to wealthy (more on all this in Chapter 6). For some of these (pain-med overdoses, Lap-Band deaths, suicide by cop), the end of life is a direct outcome of nanny-tech relief taken too far.

It's hard to track suicide rates over the past half-century, in part because record-keeping about this shame-shadowed and politically controversial subject is spotty—done by local officials in different places with different agendas or procedures. When a man is found hanged in a jail cell, or a woman crashes her car into a bridge abutment, or a teenager dies of a drug overdose, it is not always clear exactly what happened—or what was intended. Suicide doesn't always get recorded.

In the most recent 0.002 percent of the human journey for which we have records, though, there's clearly been a surge in self-termination. Julie Phillips, an associate professor of sociology at Rutgers University, who has published research on rising suicide rates, told the *New York Times* in 2013 that the actual incidence of the scourge is "vastly underreported" and that "we know we're not counting all suicides."

- In the first decade of the twenty-first century, suicide rates among Americans of age 35–64 rose by nearly 30 percent. Among middle-aged men (statistically the most vulnerable group), about 27 of every 100,000 men opted out each year. Imagine twenty-seven of the spectators at a typical University of Michigan or Alabama football game driving home and putting an end to their season. Those are small numbers compared to the numbers of people suffering from other afflictions, but I wonder if suicide is simply the most dramatic indication of societal discouragement. I doubt that of every 100,000 fifty-year-old men, 99,973 feel existentially OK. More likely, for everyone who departs with a self-inflicted act, there are many others who stop just short of that last step—and more still who pull back just a little before that.
- In the US military, we've experienced an alarming surge in suicides—to the extent that by 2012, more soldiers were dying by their own hand than were being killed in combat in Afghanistan. A Veterans Administration study estimated not long ago that twenty-two vets were committing suicide each day. Suicide, like mental illness (which is often its proximate cause) is not in itself a weakness. But as with mental illness, the conditions that produce enough pain or despair to override the human instinct to survive—those conditions *are* a weakness of society at large.

Sign 8: Shrinking Brain Parts

In millions of us, certain parts of our brains have gotten smaller. In some of us, it's the hippocampus that has atrophied; in others it's the thalamus, or amygdala, or prefrontal cortex. It took hundreds of thousands of years for the human brain to evolve to its present configuration, so if some of its parts have shrunk enough in some individuals to significantly alter the design of the whole within just the last 0.2 percent of our time so far (and possibly in just the last 0.02 percent), it may signify a destabilizing of the thing that has controlled human development. That kind of destabilizing, that sudden, in that many people, signifies a serious weakness in the population at large.

In my initial review of reports from brain-shrink studies that concern particular structures, I found one brain part for which the shrinkage was attributed directly to overdependence on a specific technology.

The hippocampus, which among other things gives us our capability to form mental maps and build navigational skills, was measured in several studies of London taxi drivers, by researchers at McGill University in Canada and the University of London. The researchers compared two groups: drivers who had relied on GPS to find destinations for at least three years, and those who'd relied only on their own memories and knowledge of the city's notorious tangle of streets. The brains of both groups were scanned and in the group that relied on GPS, the hippocampus was found to be significantly smaller. Either the hippocampus had grown larger in the self-navigating group, or in the GPS group it had shrunk. Either way, the human brain is quite "plastic" and responsive to cultural changes or events in ways that can cause parts of it to either shrink or grow.

Other studies have found shrinkage of brains to be correlated with an intriguing number of conditions of modern life—beginning with our proclivity for growing old, but also including a wide range of stresses and illnesses. Loss of brain mass with ageing is so common that it's considered normal, along with the loss of muscle mass or speed.

I don't know what to make of this phenomenon of creeping decrepitude, as I'm in the inexorable process of ageing myself. It worries me, because I know the part about declining muscular strength and speed is unfailingly true. As a long-distance runner, I can train as hard now as I did thirty or forty years ago, and with greater knowledge of training methods and physiology, yet the clock says I'm now dramatically slower. But the prospect of comparably declining cognition and insight—and memory? It seems unbelievably cruel. On the other hand, I have to remind myself that in hunter-gatherer days I'd very likely have been picked off by an infection or predator long before now. I'm lucky to be alive at all.

I made a list of the conditions that had been linked to brain shrinkage in studies at Yale University, the University of California at Davis, and the National Center for Geriatrics and Gerontology in Japan, as well as at McGill University and the University of London. They include obesity, diabetes, high blood pressure, lack of physical exercise, chronic back pain, and a range of addictions—to alcohol, cigarettes, heavy marijuana use, and video games—and to traumatic life events. There's conspicuous overlap between this list, except for the "traumatic life events," and the list of societal weaknesses described in this chapter, all of which have in turn been linked to heavy tech dependence.

As I thought about it, the idea that heavy tech reliance might cause brain shrinkage made sense in the case of the GPS taxi drivers and the hypothalamus, but also seemed in the case of the "traumatic life events" to be counterintuitive. A big part of what separates us from the hunter-gatherers is our social safety nets, and a big part of those nets is our nanny-tech protections: the helicopter rescue of a lost hiker who was too clueless to stay on the trail or carry a map; the seat belt in your car, the air bag; the wheelchair ramp at the mall; the bicycle helmet; the security camera at the teller's window; etc. If brain-shrink is correlated with trauma and stress, you'd think devices that make life safer or easier, and thus *less* traumatic or stressful, would *reduce* brain-shrink.

For a different perspective, though, it might be eye-opening to compare the number of people directly affected by trauma, for example, with the number who are exposed to the phase-3 disconnections, night after night, of TV shows that *depict* trauma—telling us how omnipresent are the threats of violent crime, terrorism, war, fatal accidents, and disease in our troubled world. *Be afraid!* Murders and kidnappings may be rampant in America, but they're a whole lot more rampant on TV and the Internet, which are our main sources of information. According to Norman Herr, a professor of Science Education at California State University, the average American child has seen 8,000 murders on TV by the time he or she has finished elementary school. And by age 18, the kid has seen 200,000 violent acts. Yet, many parents apparently shrug off this kind of trauma as relatively harmless, because it isn't "real." The British sociologist Christopher Ward describes TV as "for the most part a passive babysitter" that keeps us transfixed.

Similarly, it would be revealing to compare the number of people who complained of pain a few decades ago to the number who do today, when advertising for pain meds is constantly in our face. A series of commercials I saw in 2013, for the pain pill Aleve, implied that for someone who feels aches and pains while working at a physically demanding job, it's good practice to ingest chemical pain relief every day. While the pain pill is sometimes a needed medical aid, more often than not it is just a nanny-tech. And so is the TV itself, which may bring you good entertainment or information, but also reminds you, night after night, of what relief you can get from the pain in your head, joints, stomach, feet, gums, shoulders, back, or butt.

As for those scattered studies that find brain shrinkage associated with such factors as alcohol consumption, video game addiction, low physical activity, and back pain, all I had to do was look a little more closely. All of those conditions are associated with the massive changes we've wrought in just the latest 0.02 percent of our existence, and more specifically through the *disconnection from our nature* we've suffered during that time (recall Chapter 3).

Traumatic experiences *may* have been as common for our hunter-gatherer ancestors as for us, but for the prehistoric hunter the potent adrenal surge caused by an encounter with a snake or boar, if the hunter was sufficiently strong and alert, was probably dissipated by the fight-or-flight response that was his livelihood—whereas in the tech-protected environment of the modern world, that response is chronically frustrated. The trauma of having friends or relatives killed in a far-off war by faceless enemies allows no fight-or-flight relief, because you've been disconnected from the event and have no direct sense of its physical reality. Most of us don't engage in bar fights, or express road rage by shooting other drivers, because if you do the system will take away what little freedom you have. But the blood-chemical cost of unvented anger is heavy.

In any case, whatever uncertainties I may have felt about the significance of shrinking brain parts were thrown into a whole new light by what was for me, at least, a more surprising discovery: that the human brain *as a whole* evidently has shrunk since hunter-gatherer times—or, more to the point, since civilization began expanding our powers far faster than it has expanded our skills at managing those powers. It's also intriguing to recall that some years ago, researchers led by Fred Gage of the Salk Institute for Biological Studies found evidence that one activity modern humans can engage in that most replicates pre-civilized activity—long distance running—may *increase* the number of brain cells. Unquestionably, our persistence-hunting ancestors had to use their brains much harder to find food than people with voice-activated fast food-locating apps in their cars do.

Sign 9: Whole-Brain Shrink

The ninth sign of growing societal weakness thus comes as a message that is just about as unexpected as anything our hubristic modern minds could have imagined: *we may not be as smart as we always thought.* Despite the astonishingly powerful technologies that surround and support us and jump to our commands every second of the day and night, we may actually be dumbing down. Of course, intelligence is not necessarily proportional to brain size in the animal world, but still.

Premonitions that the wonders of the Internet age might bring unanticipated boomerangs began appearing almost as soon as the Internet did. In 1994, a collection of essays ruminating about the incipient decline of serious reading of the kind provided by print media was published in Sven Birkerts' book *The Gutenberg Elegies: The Fate of Reading in an Electronic Age*. In a 2007 essay for the *Boston Globe*, Maryanne Wolf suggested that children who are heavy users of the Internet could end up having "neither the time nor the motivation to think beneath or beyond their googled universe" (sic). Then, in the summer of 2008, as I noted earlier, *The Atlantic* published Nicholas Carr's arresting article "Is Google Making Us Stupid? What the Internet is doing to our brains".

Carr's article provoked a predictable dispute, which The *Atlantic* editors no doubt much anticipated. In some respects, his argument was an old argument that has had no resolution: there are people who read *The Atlantic* or *Boston Globe*, and maybe even a little John Irving or Thoreau . . . and then there are the proselytes of the tech world, who extol the vastly increased reach and speed of search engines and viral videos and what my college classmate Jeb Eddy, in Palo Alto, calls "the future of reading." The debate might be described as a tug-of-war between quality of thinking and quantity of information. But in one respect, Carr took it a critical step further: he argued that the Internet may have a far more pervasive impact on the human mind than any previous phenomenon, because it is performing more and more of the cognitive functions we were born with, thereby gradually replacing them. He elaborated on this theme in a subsequent book, *The Shallows: What the Internet is Doing to Our Brains*, in which he argues that Internet reading is essentially shallower than that afforded by books and other print media—undermining our skills of concentration and contemplation.

While I found myself nodding in agreement with Carr's lament about aborted concentration and contemplation, it also seemed to me that his argument does not have to depend on anyone's reverence for traditional literary culture. In fact, it's too late for that; we've seen at least two US presidents gain popularity by appearing unabashedly anti-intellectual and ungrammatical. With rare exceptions, you don't get elected in America by appearing smarter or better educated than your opponent; you get elected by being the easy-going regular guy people might like having a beer with. And the decline in education is self-reinforcing; the less well-read we are, the less we know what we're missing and the less important it seems to us to have time or skill for in-depth reading to begin with.

But never mind that loss now, because there are two new considerations to bring to this fight for the future of the American mind, beyond those deployed by Birkerts, Wolf, and Carr. The first involves going beyond the digital revolution and Internet to a much broader review of tech impacts—a "big-picture" review of our evolutionary relationship with all our inventions. The second concerns the growing body of evidence that in our cognitive skill, we humans are being both mentally rewired and, increasingly, shut down.

The big-picture review looks at the impacts not just of the digital revolution but of the Industrial Revolution before it—not just about what Google is doing to us, but what coal-fired boilers, steel smelters, alternating current, and internal-combustion engines did before the first PC was built. Human cognition was well on the way to being usurped before Google came along. As for the hard evidence of a diminished human brain, we need only look at the links between those scattered studies of shrinking brain parts and the mounting evidence of impaired cognition, including the reports linking Internet addiction to shrinking capacity for focus and reflection. Then, in the context of these findings, consider the evidence that the brains of modern humans may not be as large as they once were.

Between the beginning of civilization and today, it seems, the overall size of the human brain has apparently diminished by 15 to 20 percent. In the last 500 years (the most recent 0.02 percent of our evolution), and especially in the past 50 years (the most recent 0.002 percent), equipped with our computers, Internet, and rockets, we have conquered the earth and are eyeing other planets—and meanwhile have lost a substantial part of the organ that enabled us to achieve that conquest. Now the conquerors are being conquered. The size of the human brain peaked at about 1,500 cubic centimeters (cc) during the time of Early Modern Humans, or so-called Cro Magnon man, 20,000 to 40,000 years ago. Now look what's happened:

Early Modern Human
(after 99.4 percent of our evolution: **1,500** cc

Modern Human Brain
(after just the subsequent 0.6 percent: **1,300** cc

The most convincing confirmation of this came in 2010, when anthropologist Antoine Balzeau of the French Museum of Natural History examined the skull of a 28,000-year-old Early Modern skeleton that had been found in a cave in Dordogne, France. Using advanced imaging technology, Balzeau made an endocast showing that the brain this skull once contained had been 15–20 percent larger than the modern human brain. Other studies, cited by University of Wisconsin anthropologist John Hawks, indicate that the shrinkage since Early Modern Humans has been about 10 percent, or 150 cubic centimeters—an amount of brain about the size of a Macintosh apple—the original kind, with a lower-case "a."

When I was a kid first learning about evolution, I heard science-fiction-inspired jokes about humans eventually becoming giant heads with tiny vestigial appendages. But now, instead, the brain getting *smaller*? Doesn't that totally contradict what we know of human progress? And how could I not have heard of this? It's not that the evidence (so far) of a significant shrink is much questioned, although a prudent researcher might not want to draw any firm conclusions until a lot more Early Modern skulls have been found and measured. Meanwhile, I can only infer that even the very idea that our brains may be shrinking is a thing that the media most of us depend on for news or stimulation have very little incentive to cover. The US economy—and increasingly the world's—is heavily invested in consumer technology sales and entertainment. It is only lightly invested in serious education, environmental protection, human health, adaptation to the far-reaching ravages of global warming, preparation for the coming destruction of coastal cities, replacement of deteriorating roads, bridges, pipelines, power lines, water mains, and other costly infrastructure, and a long list of other urgent needs of the kind it will take hard use of our brains to meet.

•

Consumer technology (including all networks and devices used to produce our ever growing tsunami of entertainment, chatter, and distraction) generates much of the revenue that, through advertising and promotion, pays for and controls the major news media. Just watch how much of the advertising is for cars, fast food, and drugs. As the Romans found, before their collapse, there are benefits to be gained by keeping the populace satisfied with "bread and circus." Their formulation was later updated by Marie Antoinette's alleged comment about hungry peasants, "Let them eat cake," before *her* brain-severing demise, and more recently by the hugely profitable but debilitating American penchant for quelling anxiety with Twinkies, Snickers, and fries. There's no political or commercial profit in pointing out that people may be getting dumber.

Of course, dumbing down is a result of multiple factors—the rise of cultures that encourage the kinds of gluttonous consumption that undermine both physical and mental fitness; the health impacts of foods processed with hundreds of chemicals that have never been tested for long-term safety (the laws of most countries including the United States allow that); and the mental-health impacts of kids growing up in an overcrowded world whose adults give them little or no encouragement to wonder and explore what the purposes of life might be, other than controlling the behavior of other people and making a lot of money; and perhaps most of all, the genetics of an interrupted evolution that for good or ill no longer weeds out weakness (more about this tricky subject in Chapter 11).

While the evidence that human brains have gotten smaller is impressive, the explanations of why that has happened are less certain. One theory is that as our species has been domesticated from the wild primates we once were, and our bodies have gotten a bit smaller and less adapted to physical fighting for survival, the cranium has gotten a bit smaller in proportion—and, with it, the brain. There's some logic to this, because like terriers descended from wolves or house cats from wildcats, we are indeed less wild. And since a big part of the brain's work is running the body, a smaller modern body (Cro Magnons were bigger) might give the brain some metabolic relief.

Another theory is that warming climate favored a smaller body. It fit what I knew personally, as a long-distance runner: running generates heat, and a thinner body cools more efficiently. Christopher Stringer, a paleoanthropologist at the Natural History Museum in London, says this observation makes sense, but can't be the whole explanation because when there were earlier warmings of the climate during our evolution, the human brain continued to grow.

A similar theory is that as humans developed, along with brains getting more capable they also became more food-energy efficient, and thereby slightly smaller. Techno-optimists might point to the obvious case of computers getting smaller even as they get vastly smarter. That theory, though, is anachronistic. The human body underwent a very slow shaping to survive extremes of heat and cold, battles with predators and infections, and the arduous challenges of persistence hunting—all under conditions of unpredictable energy availability. Under such conditions, the body achieved such exceptional efficiency that it's hard to imagine how civilization, in just the latest 0.4 percent of our evolutionary time, could have made that efficiency a whole lot higher. All evidence points to the *opposite*: a fatter-than-average person is far less energy efficient than one who is lean and fit enough to outrun (over a long distance) an antelope or a horse. A person who is protected 24-7 by seatbelts, air bags, traffic signals, police, and the FDA, doesn't have to develop the skills of survival in a wild environment that was far more *biologically* complex than our civilized world is *technologically* complex.

Maybe the best explanation of what appears to be a general dumbing-down of our species over the past few millennia is that as civilization developed, human occupations became more specialized and a given individual no longer needed to have a comprehensive knowledge of his environment in order to survive. A farmer didn't have to be an expert at hunting; a soldier didn't have to be an expert at farming; a weaver didn't have to be an expert soldier. Specialization brought a wide range of support systems. No single person had to be an expert at survival. And since the demands of survival are what drove the evolution of the brain, having fewer of those demands meant the brain had less work to do. One area of it might have had to work harder than in previous generations, but other areas languished. The science fiction writer Robert Heinlein apparently sensed that, when he made this perhaps wistful observation:

> A human being should be able to change a diaper, plan an invasion, butcher a hog, design a building, write a sonnet, balance accounts, build a wall, set a bone, comfort the dying, take orders, give orders, cooperate, act alone, solve equations, analyze a new problem, pitch manure, program a computer, cook a tasty meal, fight efficiently, die gallantly. Specialization is for insects.

Except regarding the "die gallantly" embellishment, I agree. When it's my time to die, I'd prefer to just close my eyes and be gone. But the sad thing about this list is that if you look closely, except for the diaper change (that's the thing we should all do gallantly), it's actually a list of specialized occupations invented by civilization: the military commander, butcher, architect, poet, doctor, priest, soldier, etc. The real insight here is that that kind of versatility was essential to human survival before civilization began, but is unheard-of now. A Paleolithic human had to be able to take care of a baby without *having* diapers. And rather than just butcher a domesticated hog, he had to be able to track a wild and vicious boar through rough country and kill it without a gun . . . and yes, to both cooperate and act alone with a wide range of skills that the modern specialist never needs. It's a good bet that every part of the prehistoric brain got exercised hard.

Whatever the true explanation for overall brain-shrink in modern humans might be, it's worth asking this: Does it make civilized humans less intelligent? Circumstantial observations would indicate emphatically no. Civilization brought us Aristotle, Shakespeare, Madame Curie, and Albert Einstein. But of course, recent thinkers have had the huge advantage of being able to build on the thinking of hundreds of earlier people in their fields, thanks to the marvels of writing and record-keeping. Paleolithic people didn't have that—they were very busy surviving as individuals and clans with very little technology or communications from earlier generations to support them.

As I pondered this, it struck me that there's a distinction to be made between being intelligent in the conventional sense, and being a good survivor. And it might well explain why at least some modern humans have been so astonishingly creative in music, art, engineering of skyscrapers, or design of virtual worlds for entertainment—yet so clueless when it comes to collective survival.

For the Far-Seeing

Everyone has weaknesses, but our greatest focus now should be on regaining our strengths. That can look like a hopeless task, if it means trying to beat the dominant system on its own terms—trying to slug it out with Goliath. If you're struggling with depression or pain, you can't get free of your dependence on Big Pharm by upping your dosage. But you can recall David's slingshot, and find another way—to wit:

- Americans are the heaviest consumers of fitness equipment and weight-loss products in the world, but what good has that done? Most of that equipment and those products are a waste. And in any case, when the economy goes down those things won't be available. The surest way to get lean and fit is also the way that costs least and enables us be least dependent on stuff we have to pay dearly for and then usually *keep on* paying for. Instead, about all we need to regain our native strength is hard physical activity, and fresh natural food that's free of refined carbohydrates, trans fats, and chemicals.
- Getting free of that famously shrinking attention span, likewise, need cost almost nothing. Here, too, it's a matter of disengaging from tech addictions. Cut way back on time spent with electronic media (TV with its rapid-fire scene changes; social media with their Tweet-sized messages), and—so to speak—settle back into an Adirondack chair with a good book. It's those who withdraw from the rush, and re-synchronize with the slower rhythms of the world we evolved in, who will be best positioned to step aside when the rush comes to a crash.

- The same disengagements that sidestep the products of fitness and weight-loss also stem the loss of self-reliance. In dystopian America, there probably won't be anyone you can reach to repair your broken treadmill, or to repair a botched lap-band procedure. But if our fitness activity is hiking or carrying river stones to build a house (more on that later), we're practicing a kind of self-reliance we're going to need.
- One more thing, here: Where weakness persists in a person who cares deeply about the future, it should not be a barrier to attempting the quest. We can't judge, or we'll end up condemned. And if there are people we love who are afflicted by any of the troubles this chapter cites, we will nonetheless need to have them with us.

5

The New Trojan Horse

Why the Takeover Has Not Been Seen For What It Is

Our greatest collective fears, at least according to historical accounts, are the fears of invasion or conquest by malevolent or sentient forces. It's not the bombs or bullets or other weapons we fear, so much as the hostile people or aliens who wield them. Other humans bent on our destruction arouse far more apprehension in us than inanimate forces. At a gut level, to paraphrase a familiar gun-rights mantra, we're concerned about bad guys pointing guns at us, not so much about the guns they're pointing.

Part of our fears of "other" people may be culturally induced and inflamed by politically powerful demagogues—Joe McCarthy railing against Communist infiltration of Hollywood and the US State Department, or George W. Bush warning of Saddam Hussein's nefarious intentions. During the Iraq War years, American apprehensions (at least as expressed by our federal spokespersons once the war got underway) dwelled on the insurgents shooting at our boys, not on those mythic "weapons of mass destruction." If we were really afraid of nukes, for example, why was no political or major-media attention given to the enormous arsenals of hydrogen bombs stockpiled in the United States and Russia, where a warhead or two could conceivably be hijacked no less easily than five guys could hijack a Brinks money truck—or an American Airlines 757? This was one of the questions Ted Taylor agonized about. But nukes apparently don't frighten most people; the hate-fueled terrorists who might get hold of them do. A nuke is supposedly just a technological construct, free of any ill intent. When ISIS beheaded a hostage, we didn't blame the knife.

Maybe what really gets to us is what we know of our own nature—that we're all capable of impulsiveness, anger, irrational behavior, mood swings, or other potentially troublesome traits that *only sentient beings* might have. After the massacre of twenty-six children at a Newtown, Connecticut elementary school in 2012, by an unstable kid, an especially trenchant article by Anna Gorman appeared in the *Los Angeles Times* under this headline:

> When the Shooter
> Feels Too Familiar
>
> For parents of the mentally ill,
> The Connecticut massacre fueled fears

Despite the inevitable furor about guns that was reignited after that horror (and after many subsequent school shootings), the idea that it's not the guns but the shooters was what prevailed. The core message of the gun lobby—accepted with little protest by the nation—was that assault rifles aren't "out to get" innocent children. And of course, the same might be said of *any* technological device.

If you're driving a car and the brakes fail, you may be angry with your car dealer or mechanic, or with the car-company executives or their lawyers, but probably not with the brakes. You don't often hear about people taking revenge against tools or machines that have been the immediate cause of their frustration or grief. Emotionally, it's easy to blame *people*, because we know how crazy, mean, or malevolent people can be. The things people misuse? They can be dangerous if mishandled, but they don't have mood swings, and are never crazy, mean, or malevolent.

That inclination to regard tech failures with emotional neutrality, while working up righteous anger about the sentient individuals who were responsible, got a revealing demonstration in the spring of 2013, with the news of two disasters: the Boston Marathon bombing and the Texas fertilizer plant explosion. (Do you by any chance recall the Boston incident but not the Texas one?)

In Boston, on April 15, two pressure-cooker bombs exploded near the finish line of the iconic marathon, killing three spectators and injuring 260. The media reacted much as they had to the massacre of twenty-six children at the Newtown school a few months earlier. The front pages of all the major newspapers and home pages of the online media ran banner headlines; the TV networks ran continuous coverage through the evening, and the event dominated national attention for days.

Two days after the Boston Marathon, a fertilizer plant in Texas blew up in an explosion a hundred times bigger than the ones in Boston, killing fifteen people and destroying fifty homes. By any objective measure, it was a far more destructive event than the Boston attack had been in both lives and property. But it did not dominate the news and did not provoke massive speculation and commentary, as Boston had. In Boston, the suspected culprits were a couple of brothers who had emigrated to the United States and whose actions aroused speculation about possible Al Qaida connections. In Texas, the culprit (as far as was known at the time) was an inanimate chemical reaction. Industrial mishaps may occur because of human failure—incompetence or neglect or sabotage—but not because the equipment or buildings themselves were malicious. No one blamed the building. I'm guessing that people will remember the Boston Marathon bombing for a long time to come, but memories of the fertilizer disaster will fade.

Some of us feel frustrated or trapped by our machines and devices, even though we depend on them totally. My brother Bob's wife, Leslie, admits that when she was working on the IBM 650 supercomputer in 1953, she sometimes muttered an unseemly imprecation and gave the machine a hard kick. And since those days, the frustrations inflicted by our tech environment have become chronic. The damn car, the stupid hot water heater, the frigging phone. Yet, we don't really blame the gadgets; we reflexively assume the culpability lies with the people who designed, manufactured, or sold them, or maybe it lies within our own lack of skill, but not with the gadgets themselves.

One day, my daughter got exasperated with her obsolescent (over-two-year-old) cell phone and hurled it to the hardwood floor. But she then quickly picked it up and with redirected verve drove to the T-Mobile store and angrily confronted the managers—the ones she considered really responsible for her frustration. It wasn't the frigging phone that was at fault, it was the greedy people who designed it to be obsolescent in two years, and the other greedy people who parlayed that obsolescence into new sales. (The manager told her, with just a flicker of a smile, "We don't support that model any longer.")

In short, the gadgets themselves get a free pass. Smart as a smart phone may be, it isn't smart enough to have malice aforethought.

And that's where our tech-immersed consciousness has led us catastrophically astray. We correctly assume that if a gadget isn't conscious, it can't have devious intent. But we make that assumption because of a profoundly illusory perception we have of our own, human, consciousness. We tend to think that an individual person—you, or I—will have a unified identity, or self. Whatever your intentions, you are responsible for them, because society assumes that "you" have a single brain and personality.

But in recent years, the fast-growing science of neuropsychology has found that the reality isn't that simple. As we now understand, the brain is not a single intelligent entity but, rather, an assemblage of numerous separate components of intelligence. Even such a seemingly simple task as driving a car, for example, is accomplished by dividing the work into a number of smaller tasks—observing other cars or pedestrians on the road ahead, recognizing what they are, analyzing their movement, determining what response is needed, and moving the hands on the steering wheel or feet on the brakes as appropriate. The different brain parts specialize in their respective sub-tasks, smoothly communicating with each other to produce the illusion of a single consciousness.

The thousands of nanny-tech assists that surround us have an analogous existence. Each is a tiny component of something vastly larger—a de-facto capacity to collectively control human behavior. Does that vast assemblage of tiny components have consciousness? I think not. But again, remember: our own consciousness is an illusion of unified identity—just as physicists point out that our bodies give us an illusion of solidity, when our molecular structure reveals that we are actually 99.99 percent empty space. Or just as biologists point out that while we think of ourselves as single organisms, in fact each one of us is a walking megacity of millions of microorganisms, without which we would not be alive. So, while our nanny-tech environment may not have consciousness, what it does have, in its totality, is something that acts a lot like volition—the power to drive us in particular ways. As individual humans, we experience the volitions of hunger, thirst, sex, love, curiosity, fury, and fear. What the collective power of our growing nanny-tech dependence does is to drive us toward ever greater appetites for speed, simplification, and the satisfactions of the moment. It drives us *away* from the natural activity of healthy humans. It doesn't have to have malicious or evil intent to do this.

But We're Still the Boss, Aren't We?

Maybe that inanimate nature helps explain why we give our devices a free pass when it comes to blame for our weakened and passive condition. We designed these devices after some of the more mechanistic parts of our still mysteriously sentient, potentially malevolent, and deeply fallible selves—so we need to reassure ourselves that this modeling is only mechanical or electrochemical, and that in matters of ultimate responsibility, *we are still the boss.*

Why do we need that reassurance? Technological invention began in prehistory with the fashioning of tools and weapons that extended the biomechanical capabilities of hands and arms—the abilities to scratch, dig, pull, push, lift, twist, pinch, mold, peel, grasp, rub, and grip. Other mammals had long, sharp teeth and claws, so for early humans in a battle for survival, some other advantage was needed. Sharp-edged pieces of chert or flint became—over generations—tools for scraping, chipping, chopping, and spearing. In effect, the first inventions were artificial claws and teeth big enough to help level the playing field for humans.

With the rise of civilization, this mode of inventiveness expanded rapidly: wheeled carts augmented legs; levers and catapults augmented arms; telescopes augmented eyes. Eventually, libraries and then search engines augmented the memory center of the brain. And less obviously but with far-reaching consequences, the use of many small devices by specialists, to collectively do gargantuan tasks that no single device or person could do alone, replicated the anatomical division of labor among separate brain parts.

An epochal acceleration of that process of physical augmentation came with the advent of coal- and oil-fueled energy and the harnessing of steam and electric power. But again, whether consciously or intuitively, the Industrial Revolution was modeled on the processes we already carried on our feet—the metabolism of carbon-based energy:

- We humans are energized primarily by carbon-based foods—carbohydrates. Our industries, too, have been energized primarily by carbon-based fuels–hydrocarbons.
- Our food energy is derived from photosynthesis—the effect of sunlight on food crops, or on feed crops for beef and chicken, etc. Our industries' hydrocarbon energy, too, is derived from photosynthesis—the sunlight on plants that over hundreds of millennia were transformed into fossil fuels.
- Our human use of energy produces metabolic waste in the form of water (sweat and urine) and carbon dioxide (exhalation). Industrial energy use, too, produces metabolic waste from the burning of coal and oil, in the form of effluent water and carbon dioxide emissions.

By augmenting—and sometimes replacing—human physiology on a colossal scale, our eighteenth- and nineteenth-century predecessors expanded their own bodily capacities to kill, haul, harvest, and build, thousands-fold. In time, the expansion of gross muscle power moved on to the realm of the nervous system—electrochemical communications within the body that were magnified in the harnessing of electric power. The computer revolution and its spectacular digital flowerings were the culminations of that. Whole industries functioned like superhumans, as did whole networks of computers. Maybe it's no wonder that along with science fiction depicting machines that had human-like capability, we got comic books depicting humans who had fantastically expanded (machine-like) capacities. All along, the role of the individual human was—with the rare exception of a dissociative identity disorder ("multiple personality")—depicted as a single, conscious, being.

•

Back in the early days of robotics, there was much discussion of cyborgs—"cybernetic organisms" that integrated tech augmentation with human bodies. For centuries, carry-along augmentations—eyeglasses, crutches, George Washington's wooden teeth—had been routine, and it was an easy transition to actually attaching man-made objects to the body: dental implants, pacemakers, or the legendary pirate's peg leg. With further advances, we got increasingly natural-looking artificial body parts, one after another, until it seemed nearly every part of a human except the brain and nervous system could be replaced. In 2013, a British TV channel assembled a "real bionic man"—named Rex, for 'robotic exoskeleton'—to demonstrate that Steve Austin, the fictional bionic man of the eponymous TV show, was no longer entirely fictional.

Rex lacked a brain, so the implicit message of the show was that if there is someone out there who has a functioning brain but a useless body, he or she could be augmented piece by piece, from the feet up, to become a whole functioning person. The feet and ankles for Rex were provided by the Massachusetts Institute of Technology; hands and arms by Johns Hopkins University and Touch Bionics in Glasgow; heart by Syn Cardia Systems in Arizona; eyes by the University of California; spleen by Yale University—and so on. I did notice that quite a few parts weren't accounted for (no face, for example—a person really needs a face). But that needn't throw cold water on the idea, since stem cell technology will soon let us grow living parts that are even better than artificial ones.

Meanwhile, the artificial parts earned real praise. The prosthetic foot and ankle, for example, had been designed by M.I.T. professor Hugh Herr, who had lost his legs in a climbing accident and was now fitted with the same model provided to Rex. Herr's prosthetics had sensors that coordinated with his body movements as needed for walking, running, or climbing. He said he could now climb even better than before his accident.

Not everyone was happy with this show. As reported in the *Daily Mail* (UK), a professor of bioethics at Boston University, George Annas, commented, "I think when it comes to our bodies, the danger is we might change what it is to be human." Annas then raised the specter of a neo-Frankensteinian future, where "your creature let loose in the world becomes destructive and uncontrollable."

Annas may have too quickly glossed over the fact that all those artificial parts, when implanted in real people, are still under the control of those people's brains and, ultimately, their DNA. There may be nothing necessarily dehumanizing about having an artificial heart, for example. The company that made Rex's heart, Syn Cardia, has had over a thousand of its synthetic hearts implanted in human patients—with no Frankensteinian consequences as far as we know. Synthetic body parts and prosthetics might be considered an often *good* use of nanny-technology—ask that M.I.T. professor who likes to go climbing. My wife Sharon has an artificial knee, and we are grateful that she does.

But where the story gets a bit scary is in the very slippery slope from artificial body parts to artificial intelligence. Some of us might want to draw a red line between a policy that allows a guidance system in a robot that's programmed to do specific tasks, and a policy that allows artificial intelligence to be embedded in a living human's genetic makeup and memory. The former is limited to its assigned tasks. The latter is a form of Russian Roulette, in which all of us take part whether we know it or not.

Similarly, with the booming development of the Internet of Things (IOT), we might want to make a careful distinction between (1) linking us with our things, and (2) linking our things with each other, if only via other people's thumbs. God forbid that the electronics controlled by your car's accelerator pedal should somehow get connected to your Dad's pacemaker. Is there a possibility of mischief or malice here?

Your Phone is Smart, but It's Not Conscious!

With the blooming of genetics and neuroscience, boosted by computer muscle, the notion of an individual person's having a single, unified consciousness was found—as noted above—to be largely an illusion. The early psychoanalysts had raised eyebrows by theorizing that an individual is a union of multiple parts, often in conflict—but they'd had little impact on the general belief that a normal person has a single identity. Freud had theorized (based mainly on endless conversations with his patients) that every person has three separate selves—id, ego, and superego—which have very separate interests. But the theories of Freud and others of his time were later dismissed by hard scientists as unverifiable. Then came the technologies of brain scanning, and the realization that the human brain could actually be mapped. Two general features of the brain became clear: its parts are specialized, and its ability to function (and cope with the conflicts Freud described) depends on *inter*dependence—the cooperation of its specialized parts.

For me, recognizing that interdependence—the symbiosis of specialization and cooperation—was a revelation. It seemed to explain why the technologies we invent don't need even the slightest consciousness or intentionality, as individual devices, to be able to collectively take over our powers and quite mindlessly impose their functional volition on us—a volition to increase power and speed in all things. It's like the power of the ocean to erode sea walls—not with intention, but nonetheless with relentless force. If the history of civilization has been largely one of modeling our technologies after our own organisms, it stands to reason that the dispersed powers of the brain, while not capable of consciousness on their own, are uncannily and enormously replicated in the dispersed powers of our nanny-tech environment. We see that our door openers and TV remotes don't threaten us. We don't fear that the smart phone in our pocket might be a Pandora's box. What we don't really notice is the collective impact of the thousands of individually innocent nanny-techs that are replacing huge swaths of our evolved capacities.

Even if you agree that the total effect of a thousand helpful nanny-techs is to get us caught up in a world of ever quicker gratifications, speed, and fading interest in the past or future, is that necessarily *bad?* It's a legitimate question now. Maybe if you'd asked an emperor or pharaoh or the architect of a great cathedral of centuries past how important it is to plan for the future, he'd have considered your query absurd, or even contemptuous. Judging by the attention given to building temples and monuments that would last thousands of years, for at least a powerful elite the future was *huge*. Ask people that question now, and yes, some people care very much about the earth or the Seventh Generation, or at least want their children to get a college education and have a shot at a better life. But there seems to be a large majority, now, who no longer value the future as earlier generations may have. Many are conditioned by the fact that popular culture is about *right now*, and does not encourage long-range envisioning. Here's a clip from an AT&T commercial showing a man speaking to a group of children: He says:

> "Why is it better to get what you want *now*,
> instead of later? . . .
> It's not complicated. Now is better!"

And here's a recent advertising pitch from Pepsi:

> Live for the moment!

The authors of the American Constitution endeavored to look far into the future of their country, and wrote a guide that still seems to be regarded by most citizens as largely unquestioned authority two and a half centuries later. But I know of nothing comparable to that effort in recent years—even though the changes confronting us now are far greater than the changes the founding fathers were confronting in their break from authoritarian forms of government. Now we're facing a break from past forms of human life itself. You'd think our think-tank wonks and legislators would at least propose that we write a new Constitution for the age of global population pressure, climate disruption, ecological deterioration, technological addiction, and rising mental stress, but they have been defeated by a Trojan Horse they never saw coming. And when it came, everyone thought it was just more horsepower.

There's a smaller number of people—a kind who rarely run for public office—who *do* envision, and who have become darkly pessimistic or even nihilistic about what they foresee. I've heard at least four prominent scientists say, in private, "we're finished." Most of them can't say this in public, or they'll be dragged out of town chained to the rear of someone's pickup truck. But most climate scientists at least agree that we've gone well past the point where severe climate disruptions can be averted. And "severe" is euphemistic. In 2015, the

Bulletin of the Atomic Scientists re-set its "Doomsday Clock" at three minutes to midnight—leading *Time*.com to summarize: "Climate change and nuclear proliferation make global catastrophe highly probable, scientists say."

The focus of US public policy, having missed the boat on global-warming mitigation, is shifting grudgingly to adaptation. A few years ago, some urban officials began wondering whether sea-level rise might actually do serious damage to coastal cities. Now, for those who do look ahead ("the far-seeing"), the more pertinent question concerns which parts of which cities can be saved, and how. But most officials evidently don't even wonder that, because the nanny-tech usurpation has numbed or shut down their envisioning capacity. Their days are filled with speed-dialing campaign contributors, texting with staff to put out political fires, tweeting to constituents, preparing sound bites for TV . . . while outside the building, the temperature continues to rise. Some of us aren't going to just wait.

For the media that control the bulk of what most of us know, there's motive for averting eyes from even the most urgent threats to the world, if the coverage of those threats is unsettling to TV or web viewers (who will quickly turn to something else) and thus suppresses ratings and revenue. One result—deeply ironic, in this age of growing speed in nearly all things—is long delays in acknowledgment of what's really happening. In late 2013, the IPCC issued an updated assessment finding that sea level was now rising much more rapidly than at the time of its previous reports, and that the rise is accelerating. According to the report, by 2300 (less than the time between the writing of the Constitution and the writing of the first IPCC report), the Atlantic Ocean will have risen by up to three meters—about ten feet. Envision a hurricane more powerful than Superstorm Sandy hitting New York from a base that's ten feet higher. New York City—along with huge areas of the entire East Coast—could be mostly gone. And significantly, the most recent IPCC report says the rise is now irreversible.

For some regions of the oceans, due to variations in thermal expansion, the news is even more alarming: just two years after the IPCC report, in 2015, researchers at the University of Arizona and the National Oceanic and Atmospheric Administration (NOAA) confirmed the IPCC's warning about accelerating rates of increase by announcing that the sea level along the northeast Atlantic coast of the United States had risen by nearly four inches in just two years—a rate of increase which if continued would lift the sea ten feet in just thirty years. Whether or not that rate continues, that finding doesn't bode well for Portland, Boston, Providence, or New York.

Regional variations aside, the IPCC's world report was one of the most momentous pieces of news about the human prospect yet, and I watched for it on the network broadcasts. Had journalists (or their employers) learned anything since the first IPCC Assessment and the *World Scientists' Warning?* There was a lot about shootings, political scandals, and bus crashes, but I saw nothing about what was happening in the skies. Too much of the news industry's revenue was supplied by advertising for fossil fuel industries—cars, oil, coal, natural gas, petrochemicals, and coal- or gas-fueled electric power, as well as for electronic gadgets, games, and entertainments that keep us reflexively spending, passively consuming, and obliviously dumbing down.

A study of how well-informed the American public is (or is not), conducted by political science researchers at Fairleigh Dickinson University, provoked this headline in *Huffington Post*:

> **Fox News Viewers Know Less Than People Who Don't Watch Any News**

Aside from the not-so-funny question of whether it's possible to know less than nothing, the story may have added to the mounting uneasiness, among many, about the magnitude of public ignorance. Data about this painful subject vary, but the conclusions are usually bleak. A widely cited study, albeit hard to believe, found that 42 percent of college graduates never read another book after college.

Whatever the real numbers, we may have come to a place where for the many, it only dimly matters what tomorrow holds. That may be especially true in places where much of the human environment no longer consists of other life forms but of inanimate toys (both children's and adults'), and where the toy-makers and marketers are unobtrusive. The business of advertising is very astute about giving a non-threatening human face to its products. Notice that the characters who appear in commercials for Apple, AT&T, and Time-Warner are mostly goofy, nerdy, nice guys. Harmless, funny, congenial—nothing like the malevolent, scheming, corporate criminals or sociopaths we meet in the detective shows or movies.

A memorable example was the "Get a Mac" series of sixty-six TV ads for Macintosh computers which appeared for several years featuring comical encounters between the bumbling man named PC and the hip, unflappable, slightly younger-looking guy named Mac. PC, the personification of an IBM personal computer, is a bit overweight, stodgy, and uptight. Always dressed in a suit and tie, he's formal and good with numbers but prone to pratfalls—a congenial doofus. Mac, on the other hand, is lean, casually dressed (usually in jeans)—a cool, good-humored guy who seems a bit amused by his rival's ineptness. Neither character is overtly hostile or aggressive; they're just a couple of regular guys. Their encounters were always fun, and the campaign was a huge success for Apple. If the *people* you associate with a tech product seem like nice guys you'd get a kick out of meeting and having a beer with, you're not going to feel threatened by the product they represent.

How Science Fiction Misguided Us

The first man-made things to exhibit a form of arguably human-like intelligence were the giant scientific and commercial supercomputers built by UNIVAC and IBM in the early 1950s. Those car-sized machines were as challenging to our ideas about the powers of technology as had been the atomic bomb a decade earlier. It's easy to see why the science fiction stories of that era began to depict such machines as techno monsters—capable of rivaling and seizing human powers.

The most famous of these post-Frankensteinian inventions was Hal 6,000, the emotionless usurper in the 1968 movie "2001: a Space Odyssey." But by the arrival of the actual year 2001, the specter of such usurpation was no longer quite so titillating. Computers were no longer alien. In the half-century since their arrival, their development had taken two quite disarming directions: first a scattering of tasks (and powers) among hundreds of thousands of small devices so that no single device seemed to pose any threat to its user's autonomy; and second, of course, extreme miniaturization.

To put the miniaturization we've witnessed in its proper perspective, you have to consider not just the relative sizes of new computers versus old, but—more significantly—the relative sizes per gram of computing memory and power. The original UNIVAC 1 weighed 29,000 pounds (13 metric tons), and my sister-in-law Leslie recalls that she found it mind-boggling how fast it could do a thousand calculations. One day, decades later, I asked her how long it would have taken that machine to do . . . not a thousand or million or billion, but a *trillion* calculations. She recalled the machine's speed, then we figured that if it had been run day and night, and kept from overheating, a trillion calculations would have taken the machine around fifteen years. Now we can do that in less than a second. And the speeds at which we can do other tasks have increased commensurately. Yet, the new speeds are barely noticed, because those gigantic powers are dispersed among thousands of uses, and even those uses are barely noticed—or are hidden from public view.

So, it's not just that nanny-tech assists get a free pass because they show no malevolence. Even if they had intentionality, and assuming we didn't get too careless with the IOT, they'd be specialized for small tasks and too cut off from big-picture consciousness to act—or even think of acting—contrary to their designated job descriptions. In a world beset by sweeping climate change, failing health, and a host of other dangers, our little tech devices pose no visible threat. On the contrary, for a lot of us in a chaotic world, they may be the one aspect of our existence that's most predictable and reliable. They are our dependable nannies. And because our vigilance has been elsewhere, they have taken control of our lives without a shot being fired.

There are historical precedents for invisible invasions that brought powerful civilizations to their knees. The unseen invasion—the conquest not recognized until it is over—is immortalized in the legend of the Trojan Horse, which instead of being battled by defenders is brought into the city by the people being invaded. That story is apocryphal, but it sticks around in our lore as a reminder of how easily ambushed people can be. When Europeans came to the Americas, it wasn't their swords and guns that decimated the Indians as much as the microorganisms they carried—unknown to the natives who'd never developed resistance to them.

Humans have always been vulnerable to invasions that quietly circumvent any awareness of danger. Biology students peering through their microscopes at tiny organisms have occasionally marveled at how similar they look to the giant monsters or space aliens depicted in movies. Is it from the photos of those organisms that the filmmakers got some of their ideas? If it's true that the technologies we build for our expanded power and convenience are fundamentally modeled on our own anatomy, maybe it's not surprising that some of the greatest *imagined* dangers to our wellbeing—our fictional monsters and body-snatchers—have been modeled, whether consciously or not, on some of the most unseen dangers to that anatomy.

We now have a vast medical industry based largely on our ability to recognize and fight the once-invisible threats to us from microorganisms, yet we have no comparable industry defending us against today's equally invisible usurpations by micro-technology. We have big governmental and public-service bureaucracies warning us of the dangers of junk food, but no-one warning us about the risks of using a fast-food delivery system that employs hundreds of passivity-inducing industrial processes to spare you the effort of shopping, cooking, loading the dishwasher, and taking out the garbage. Now you can use the drive-through. It's a double-whammy: the processed food goes to your belly or thighs, and the delivery system (car and pickup window) eliminates any physical or mental stimulation you'd have gotten from more traditional food-preparation activities.

There was a time when if you wanted a strawberry shake, you might have to go out to pick fresh strawberries, milk a cow, use your skill and wits to steal wild honey from a hive, mix with a hand crank, freeze in the ice box, and mix again. Now you can just stop your GPS-guided car at a drive-in's voice box and say: "Strawberry shake." And when you're finished, you can toss the big plastic cup, which can later become part of a giant floating island of garbage in the ocean. No need to wash anything. You feel fortunate to live at a time when you don't have to do all that business with the cow and the bee stings, etc. But you probably don't know one-tenth of what the farmer knew about the nature of the living world you depend on, or one-hundredth of what the hunter-gatherer knew. You may not even know much of what your great grandparents knew, about the rewards of getting together with family or friends to prepare a meal from fresh ingredients.

The Question We Don't Dare Ask

The question of why we haven't seen the nanny-tech takeover coming has more than one answer, and one of them lies outside the familiar territory of who profits from not alerting the public, or how easy it is to dismiss any suggestion that a takeover has really happened. Some things are simply too difficult or too unthinkable for most people to acknowledge.

Atul Gawande, a surgeon with the Harvard School of Public Health, has pondered the question of why it is that while some important new discoveries stimulate revolutionary changes, others take an exasperatingly long time to be accepted—and the difference can't always be explained by who profits from those discoveries or how proven they are. As an example he cites the difference between the rapid adoption of anesthesia for surgery, and the long ignored or slow-dawning recognition of antiseptics to kill germs. In a *New Yorker* article, he identifies several factors that might explain the difference, a key factor being that the benefits of anesthetics were immediate and dramatic (screaming patients replaced by quiet, painless operations and a less stressful environment for the surgeon), whereas the benefits of antiseptic measures were not immediately visible. Doctors couldn't actually see any germs causing infections, and infections might not appear until hours later, with no sure link to any particular cause, such as the doctor having failed to wash his hands. Gawande also mentions, in passing, other major cases of such invisibility—notably, global warming—blocking timely recognition by society at large.

It struck me, though, that Gawande might himself be blinkered by an invisibility problem at another level. In his article, and in his work for the World Health Organization, he understandably focuses his attention on the practices and technologies that can save lives and that, when not used, allow unnecessary deaths. He emphasizes that saving lives is not just something that happens in the operating room; it entails post-op care and what happens after the patient leaves the hospital. It may involve what is done or not done for years. But the bottom line is whether the person who was treated dies too soon.

There's a thing that Gawande doesn't discuss. Using technology to save lives, in circumstances where prehistoric humans would have died, overrides evolution and in some respects *improves* on evolution. We didn't evolve with our bodies able to release antibiotics far more powerful than our normal immune systems. So far, so good. What evolution *did* do, though, was to provide a more brutal means of strengthening our species' capability to survive—by sacrificing our weakest individuals, those with illnesses, injuries, or disabilities rendering them ill-equipped for surviving or having children. We now do the *opposite* of that; we do everything possible to keep everyone alive. Biological evolution may be making a belated comeback by doing things like reducing male sperm counts or introducing new diseases, but our sprint economy (Chapter 6) can outrun evolution, albeit just for a very short distance. For nearly every one of the nine categories of widespread societal weakness cited in Chapter 4, we now have an industry, charity, or government agency dedicated to protecting the people who have that weakness from being killed by it. And one of the long-term effects may be that the weakness is perpetuated.

Before this is over, I'll have to raise the question of whether overriding one part of evolution while letting related parts go untouched might unbalance the whole system and cause long-term impacts far outweighing the short-term benefits. Saving thousands of lives today, and even more tomorrow, keeps population growing at a higher rate than it would otherwise, which will probably result in the loss of far larger numbers of lives in the future.

That might seem to pose a dilemma for a doctor like Gawande, whose work is particularly focused on saving lives in India, where about 1.3 billion people (as I write) contribute more to global population growth (or "demographic momentum") than the population of any other country including its fellow giant China. And Gawande is far from insensitive to the unwanted burden of future consequences carried by health professionals in the present moment. He tells of meeting a nurse who is very competent but who unaccountably overlooks one of the time-honored practices (warming a newborn by direct contact with the mother's skin) that can affect mortality later on. "Everything about the life the nurse leads—the hours she puts in, the circumstances she endures, the satisfactions she takes in her abilities—shows that she cares. But hypothermia, like the germs that [Joseph] Lister wanted surgeons to battle, is invisible to her."

To most policymakers, it seems, the dilemma of "discounting the future" is shrugged off. To a politician, the future is next month's tally of campaign contributions, or next year's election. In next year's election, the citizens of a century from now have no vote, even though their lives may be even more at stake than ours. To some bioethicists or futurists, though, the dilemma—saving thousands of lives now may cost larger numbers of lives in the future—weighs more heavily. It's a conundrum that anyone who has ever had a major operation under sterile conditions, or who has a family member or friend who has had such an operation, would be unlikely to want to even think about. And so most of us *don't* think about it. And if we don't think about something as existentially troubling as that, we are even less likely to think about the cumulative long-term effects of technologies that seem no worse than trivial. Nanny-tech devices are so much *of the moment* that we're just not interested in how they might affect us tomorrow, say nothing of half a century or seven generations from now.

One last note: while I've focused in this chapter on "why we haven't seen the takeover for what it is," some critics might wonder about a far more talked-about, impending takeover—that point in time when machine intelligence exceeds human intelligence and the takeover is irreversible. Another view, though, is that the whole notion of the Singularity is based on an artificially narrow definition of what human intelligence means. Our invented technologies are inherently unlikely ever to exceed the full-spectrum intelligence of an organism that has gone through two-and-half million years of on-the-ground testing, redesigning, and debugging. Computer capacity is amazing with numbers, but doesn't scratch the surface of evolved capacities for wisdom or wit, sense of tragedy, or ability to envision.

On the other hand, in the realm of the sprint economy (next chapter), we have *already* hit a technological wall, and that wall now completely surrounds us. It cuts us off from our origins, our history, and each other, even as it lets us delude ourselves with the idea that we are more connected than we've ever been before. It has none of the momentous look or feel of an epochal Singularity, and that's just what makes it so hard to even notice.

For the Far-Seeing

To free ourselves from the clutches of the industrial behemoth we were clever enough to create, we need first to disabuse ourselves of the belief—drummed into us by our cultural managers—that technology is mindless and therefore morally innocent.

Yes, I do recall, elsewhere in this account, how the bomb-builder Ted Taylor and my brother Bob and I wrote a mission statement that began "Technology is neither good nor bad. . . ." Well, we were all younger then. What we didn't give sufficient weight to, then (although Taylor gave it great weight in his journal), is that the *availability* of a tech, not its admittedly non-existent good or bad intentions, is what most heavily determines the moral outcome. If there were no addictive drugs in Detroit, there would be no drug addicts there. Or for a bit more realism, Detroit was once a thriving city because it had a thriving motor vehicle manufacturing industry—and many thousands of tech jobs—but when the "big three" auto makers (GM, Ford, and Chrysler) began to outsource the bulk of their manufacturing jobs, the city began to die.

Economic research in the past few decades has made it clear that the availability of a capacity—a product or tech that augments human powers or pleasures—nearly always increases the incentives to use it. The proof is in the markets for new techs, for which no demand existed before their invention, but for which demand boomed once they were available—witness the demand for Facebook, flat-screen TVs, the latest i-phones, YouTube, or the next big thing. Amazon, the richest company in America, is in the *business* of creating demand where none existed before.

So, it's a mistake to give our techs a moral free pass. The mantra that "guns don't kill people, people do" is disingenuous, since many of those who have been killed by gunfire would not have died if the weapon used had been only a fist. There was a time when most things were made only because they were needed. Now, a large share of what's produced is sold to consumers who never wanted it until after it was invented and advertised. If a new, post-collapse civilization is to be sustainable in a way the old one is not, it will need to be free of artificial markets that create desires and dependencies where none existed before.

As we form our far-seeing communities and move forward, there will be strong temptations to establish laws banning or restricting the use of technologies that weaken or replace our natural strengths. Succumbing to those temptations, too, might be a mistake. Laws are shaped by society, but it should not be the other way around. Our strengths are in healthy individuals and societies, and when we are healthy our demands—and the markets—for destructive or dangerous techs will abate. The humanity of the future may not *need* laws banning assault weapons, addictive drugs, and nuclear bombs if there is no demand for them.

And so it comes down to people—*us*—after all. But only if we all participate in thinking it through (not letting our dominant institutions tell us what to think), and only if we continue to think it through for as long as we can breathe.

6

The Stumbling
A Slow-Footed Species in a Sprint Economy

"Many believe that the progress of mankind depends on ceaseless technological improvements measured by such elements as speed."
–Charles A. Lindbergh, *The New York Times*, July 27, 1972

"A test flight of an experimental aircraft traveling at 20 times 'the speed of sound ended prematurely Thursday morning when the arrowhead-shaped vehicle failed. . . "
–*Los Angeles Times*, August 12, 2011

"New fast-acting Advil. . . . It's built for speed."
–TV commercial, November 9, 2013

"It's our misuse [of technology] that is the problem, and the misuse of it being programmed for speed and not content. But more than that, this wonderful technology, that comes from the fantastic brains that God has given us, is now actively coming between us and God—and that is a very, very serious issue for me."
–Archibald Hart, Senior Professor
Fuller Theological Seminary, School of Psychology

I see a great irony in our twenty-first century pursuit of ever greater speed and power in all things: it suppresses our own evolved best nature. Speed and power were never our biological strengths as we evolved. If anything, they are even less so now. Compared to the powers of the inventions that now surround and assist us, we look—and must feel, if we admit it—increasingly weak and out of breath.

Early humans had to survive in a world of wild animals that were quicker, more powerful predators than they. How did those naked ancestors of ours survive without being chased down and devoured by lions or gored by rhinos before they could reach the age of thirteen or fourteen and reproduce? Bearing in mind that they had none of the routine technologies of protection that we have—no rifles, binoculars, cell phones, four-wheel drive vehicles, Kevlar vests, EMTs standing by, or even antiseptics and first-aid kits—how could they be both successful at avoiding predators and able to eat big-game meat themselves?

In 1999, when I wrote my *Time* magazine essay "Will We Still Heat Meat?" (Chapter 2), my answer was: "Maybe not, if we wake up to what mass production of animal flesh is doing to our health—and the planet's." I was aware, by then, of the emerging persistence-hunting theory of human evolution—that early humans gained a foothold on the planet by chasing down and killing large animals for their meat. And I surmised that a century or two from now, there may still be a few remaining wild places where small numbers of people hunt deer or raise goats or chickens. But for a global population projected to be nine to ten billion within the next few decades, the ecological cost of mass-producing meat as we do now is prohibitive.

Ecologists have estimated the long-term carrying capacity of the earth to be more like four or five billion people—and even at that size our species is putting enormous strain on the planet's fossil energy and fresh water systems. According to Cornell University ecologist David Pimintel, it takes about eight times as much fossil fuel to produce a pound of meat as it takes to produce a pound of plant protein. And with fresh water, the disproportion is even worse: more than half of all the water used for all purposes in the United States is used for livestock production. Pass up one hamburger, and you'll save as much water as you save by taking forty showers with a low-flow nozzle.

According to the World Bank, in at least eighty countries the supply of fresh water is no longer meeting demand. In the years to come, nations may be fighting wars over water as aggressively as they've fought over oil. So, meat's days are numbered. But the fact is that early humans *did* eat meat, and therefore did have the physical capability to catch other animals. Early humans, while physically weaker and slower than deer, horses, antelope, mammoths, elephants, and lions, nonetheless killed and ate all of those animals—even though they had no guns. By the time *Homo sapiens* emerged, our earlier ancestors had learned to make and throw spears, but try to imagine yourself facing a charging lion or elephant with just a spear. How was it possible?

Until the late twentieth century, scientists assumed the survival of early humans in a world of dangerous mega-mammals could be attributed to a gift of god or nature—namely their bigger brains. The idea was that somehow, humans had compensated for their physical shortcomings by *outsmarting* both their predators and their prey. But that facile idea begged the question of how people got so smart in the first place. How did *Homo erectus* or *Homo habilis* survive being mauled by tigers and bears for tens of thousands of years *before* their descendants got smart enough to make guns?

In the 1980s, biologists at the University of Utah and University of Vermont, soon to be joined by colleagues at Harvard and the Sports Science Institute of South Africa, found an answer that was revolutionary and compelling: a key to human survival, they found, had been the hominins' ability to physically chase down those bigger animals for food. It was crazily counterintuitive, because in the world of large mammals, humans are notoriously slow runners. How on earth could running ability be an evolutionary advantage? The answer lay in a distinction that had rarely been thought much about, or even much noticed, in our speed-infatuated culture. Physiologically, for us bipeds, there are two very different kinds of running: anaerobic and aerobic. *Sprinting*—the kind of all-out running you can keep up for a minute or two at most—is anaerobic. Long-distance running is aerobic. The Utah scientists hypothesized that while other large mammals rely on their short-distance speed to kill prey or escape predators, early humans relied on their endurance.

The story of how the persistence-hunting or "running man" theory of evolution came about is now legendary: in the early 1980s, a University of Utah graduate student in biology, David Carrier, suggested it to his professor, Dennis Bramble, who instead of quickly dismissing it as the kind of wild idea you might expect from a kid, regarded it as an intriguing insight—worthy of further consideration. Meanwhile, at the University of Vermont, another professor of biology, Bernd Heinrich, was coming to a similar conclusion. Heinrich was a national champion ultramarathon runner, and knew from his own experience that a highly trained man could run for many hours without stopping to rest—which most animals could not. Heinrich may also have had his interest in the volition of survival piqued by his experience in Germany in the 1940s, when he and his family had to literally run for their lives, first from the Nazis and later from the invading Red Army.

Carrier, Bramble, and Heinrich were later joined in their radical reconsideration of human origins by Daniel Lieberman, a professor of evolutionary biology at Harvard. In November, 2004, substantiated by anatomical as well as anthropological research, the running-man theory was featured as a cover story ("How Running Made Us Human") in the scientific journal *Nature*, and subsequently the old idea that we humans came to dominate the earth because of our big brains was dismissed for good. Rather, it was our unique strategy for survival, and the evolving skills that that strategy required, that made the brain grow. Here's Utah's David Carrier, after many more years of corroborating investigation:

> The factor that differentiates hominids [early humans] from other primates is not large brain size, but the characteristics associated with erect bipedal posture and a striding gait.

Of course, the kind of long-distance, aerobic running the evolutionary biologists are studying is very different from what most Americans may think of when the idea of "running" comes up—although that is changing with the booming popularity of 10Ks and marathons. In popular culture, entertainment, and daily news, "running" has long been about going as fast as you can—running to catch a bus, score a touchdown, make a fast-break dunk, chase a bad guy, escape from a building in which a time-bomb is about to explode. Running as fast as possible (sprinting) is physiologically anaerobic, bypassing the uptake of oxygen and building oxygen debt for the one or two minutes you can keep it up before coming to a gasping halt. A very athletic human sprinting full speed could never outrun a lion or bear. But in our sports and movies, it's the mega-mammalian sprint that we aspire to. Maybe, like the skinny kid who got bullied in middle school, our species is envious of power (which is linked to speed) and has become obsessed with it. And a lot of the people designing our new technologies and entertainments, especially, seem to be aware of that.

I got my first taste of that speed-envy at age fourteen, when I was a skinny ninth-grader at Roosevelt Junior High School in Westfield, New Jersey. One day a notice came around to the school from the Union County higher-ups, announcing that in conjunction with the county high school track championship that spring, there would be an event for ninth-graders—an 880-yard (half-mile) relay race. For each school, four runners would each race a leg of 220 yards. My school announced that it would hold a tryout for any boys (no girls allowed in those days) who might want to bring glory to RJHS. For reasons, I wouldn't have been able to articulate, I decided this was for me. I truly believed I was born to run. On the day of the tryout, about twenty boys showed up.

The athletic director lined us up and gave us instructions: we would sprint 220 yards across a field and a parking lot to a finish line of white lime. The first four finishers would make the team. The A.D. even had a starting gun. He raised it and called, "Ready . . . set. . . ." *bang!* I ran my heart out—and finished approximately last, gasping for breath, my lungs on fire.

The following year, at Westfield Senior High, I tried a different distance—the 2.5 miles of our school's cross-country course—and discovered that hey, I *was* born to run! And as I would eventually learn, so are we all—although many people have evidently lost touch with that genetic heritage. Some (like the four guys who made that relay team) are endowed with abundant fast-twitch muscle fibers and are good (compared with other humans) at anaerobic sprinting. (The top two runners on that relay team would go on to be star running backs on the Westfield High School football team.) But most of us, if we train and reconnect with our innate capability without getting injured, can run *aerobically* for distances that no lion or horse or giraffe or even wolf can. In elaborating on that point for *Nature*, Harvard's Lieberman pointed out that horses running long distances average about six meters per second, which is "slower than a top-class human runner."

Long-distance aerobic running, of course, is much slower than sprinting—slow enough to allow the replenishment of oxygen to keep up with its utilization by the muscles. A trained runner who keeps the pace slow enough to maintain oxygen equilibrium can keep going much longer—for some, hundreds of times longer—than a sprinter at top speed. Once, more than three decades after my gasping 220-yarder (the only one I ever tried), I did a 146-mile run across Death Valley and on up to the peak of Mt. Whitney—keeping it up for more than thirty hours, or about a thousand times as long as the one minute or so that anyone can sprint. The occasion was one of the early runnings of the annual Badwater ultramarathon, which ends at the top of the Mt. Whitney Portal, 135 miles from the start. But some of us made a brief stop at the finish, then continued on another eleven miles up the trail to the peak—the highest point in the contiguous forty-eight states. I finished eighth among the twenty runners who completed the course that year, and it was a memorable demonstration, for me, of *why* humans can run so long. The race was in midsummer, a seemingly insane time to run a hundred miles across the hottest place on earth—around 121 degrees, as I recall—and no shade.

Other trained runners, both male and female, have run much longer and farther than we did that day—and even in comparable heat. At the *Marathon des Sables* in Morocco, for example, runners go for six days on the Sahara Desert—and have to carry their own water for each day's stage. It's what hominins did on the African savanna. Bipeds with bare skin can cool their bodies far more efficiently than quadrupeds covered with hair or fur. And that, say the evolutionary scientists, was the secret of the early human hunter-gatherers. By running in a group (the archetypal hunting party), they spooked the bigger animal, which because of its anatomical differences could sprint only a short distance before overheating and having to stop.

As the animal caught its breath, panting, the slow-running, aerobic-breathing hunters, maintaining oxygen equilibrium, would begin to catch up, and the animal might get to its feet and dash away. But the hominids had the advantage of being able to blow off heat by the evaporation of sweat from their bare skin and abundant sweat glands; by the more efficient radiation from their thinner bodies and therefore higher surface-to-volume ratios; and by the cooling effect of convection, as the full lengths of their upright bodies were exposed to the light breeze they made by their movement (whereas the quadruped with its belly facing the ground got less of this effect). Their secret of superior cooling allowed the hunters to keep up their strength even as the bigger animal weakened and, after a few more rounds of getting to its feet and running off, would be too overheated or exhausted to continue. Then the hunters, with strength in numbers and with rocks or spears, would switch from aerobic to anaerobic, sprint in, and make the kill. Today, maybe it's a subconscious, primordial memory of that huge moment that can make us jump up from our chairs and shout when we watch a linebacker sack the quarterback. What the sprint culture has allowed us to forget is that the chase may have taken arduous hours before that switch to the anaerobic climax. Today, it seems, we're conditioned to expect big play after big play: sacks, interceptions, hail-Mary catches, *scores*—life as a highlights film.

The theory developed by Bramble and colleagues was a radically new conception of human evolution, but as a matter of physiological credibility it wasn't just a theory. At least three indigenous peoples—the San Bushmen of the Kalahari Desert in Botswana, the Maasai Warriors of Tanzania, and the Tarahumara Indians of the *Baranca del Cobre* (Copper Canyon) in Mexico still live that way (or in the case of the Maasai, practice it ritually—man against lion).

The evidence now strongly confirms that it was in the gradual development of this unique hunting strategy that humans developed bigger brains. Long-distance persistence hunting required hours of tracking and chasing of animals that were out of sight for many minutes or even hours. *What happened during those hours is the untold story of mankind*—or, perhaps, the story that's just beginning to be told. In recent research following up the London taxi study mentioned earlier (finding that heavy reliance on GPS correlated with a smaller hippocampus), scientists at University College London, using magnetic scanners to study the brains of drivers as they navigated London's labyrinth of streets, found that the drivers relying on their own brains were employing a highly complex "sat nav"-like system engaging multiple brain parts. The hippocampus was engaged in initial route planning; the medial prefrontal cortex tracked the distance to a destination; the right lateral prefrontal cortex registered unexpected features such as blocked intersections; the anterior prefrontal cortex was engaged in spontaneous route planning. All of these capabilities were developed thousands of years before the first GPS was ever made.

Today, we too easily assume the life of the hunter-gatherer was simple, compared to ours. On closer examination, it must have been, by today's standards, wildly complex. If the prey was faster than the hunter, as it usually was, it might have dashed out of sight long enough for most other pursuing predators to give up. For humans to keep chasing a prey that wasn't continuously visible to the eye, it would have been necessary to deploy not only highly developed tracking skills, but also the kind of patience and endurance that a lot of modern humans have lost. To keep using those skills through hours of lost contact and fatigue, the hunter would also need to have some kind of mental image to stand in for the prey until it was brought physically back into sight. That mental image—imagination, or visualization—might have been the work of the occipital cortex ("seeing" the animal in the mind's eye) or motor cortex ("feeling" the thrust of the spear yet to be made). For a species that spent a good part of its day visualizing a kill that was many minutes or even hours away, the cortex had to grow in capacity as surely as the quadriceps did.

Sprint Culture: The Anaerobic Paradigm

When civilization began, and the inventing and use of implements for farming, construction, and defense became keys to success, it was of course thrilling and revealing to find that some of those inventions—the first big expansion of what we now call technology—could hugely magnify the power and speed of the things we'd once done directly with hands and feet, eyes and ears. But in our pleasure with those magnifications we overlooked something that now haunts us: we are unable to commensurately increase the rhythms of our own physiology and neurology (our original nature), because those are intricately meshed with the rhythms of the natural world we depend on for every breath and step we take. The result: the more dependent on (and obsessed with) power- and speed-magnifying technology we become, the more out of sync we are with our own nature:

- We can use our cars and planes to increase our speed of mobility a hundred-fold and beyond. But with very limited exceptions, we can't increase our heart rate, oxygen uptake, metabolism, or speed of sleeping, even *one*-fold. In one of the limited exceptions, a highly trained endurance athlete might temporarily (while running or bicycling) increase his heart rate two- to four-fold above its resting rate. For example, the night before a road race, an elite marathoner might sleep with her heart resting at 49 bpm (a very fit person will have a slower-than-average resting rate), then during the climactic last 400 meters of the two-and-half-hour race go into an anaerobic sprint at 190 bpm. But that's close to the maximum increase any human can very briefly achieve. It takes months of training to achieve that two- to four-fold increase for a minute or two, and even the greatest endurance athletes in the world have to let go and return to resting rate after a matter of hours. The vastly larger disparities that prevail between the speeds of our tech-extensions and even the fastest-possible rhythms of our bodies and brains are an omnipresent stress.
- A huge industry of food processing and marketing technologies has gotten us to consume fast food, but we can't force fast digestion or healing or growth. After we do digest, we can flush away the metabolic waste in a couple of seconds, but can't speed up the months or years it takes for organic waste to decompose and replenish the earth—if in fact it is allowed to do so at all. Whether we're talking toilets, internal-combustion car engines, smokestacks, livestock feed lots, the waste dumped by ships, or the mountains of garbage we call landfills, the waste we dump is piled up far faster than our planet can recycle it. Why? For the first 99.6 percent of our evolutionary time, the human population was quite stable, probably at no more than a few million people worldwide. That would mean that in the most recent 0.4 percent, it has grown by around a thousand-fold—while the earth which sustains it hasn't grown at all. And about 75 percent of that expansion has been in the past 0.02 percent of our time on this planet—what Paul Ehrlich aptly describes as an explosion. So, if you count population growth alone, the human waste impact has increased *at least* a thousand-fold, while the earth's rate of waste recycling has not increased at all.

- But of course, during that time the amount of waste *per person* has increased greatly. When I drive my car (I'm still stuck with an internal-combustion engine), my tailpipe is releasing an amount of waste carbon dioxide and other pollutants in one minute that it may have taken the earth thousands of minutes of compression and fossilization of ancient ferns or trees to sequester. And it won't help much if the car is electric, because then the waste is being pumped out at the electric power plant instead of the tailpipe, but in most cases it's still CO_2—and if the power plant fuel is coal, the particulates and other pollutants are even worse than those from a car. The efficiency of energy generation might be slightly improved, but only slightly.
- Of course, that "thousands of minutes" it may have taken the earth to produce a minute's worth of fuel is only rhetoric. Over how much land? In what climate? We don't have much data. But the huge magnitude of the disparity is unquestionable, and similar magnitudes apply to our consumption of resources in almost everything we do in the modern world. If per-capita consumption has indeed increased a thousand-fold since Neolithic times, then multiplied by the thousand-fold increase in population, our civilization's impact on the earth may be a million times what it was until quite recently. In short, we are spending down that ultimate nest-egg environmental economists call "natural capital" and replacing it with waste at a rate that will soon render us ecologically bankrupt. Drive past any industrial brownfield or urban slum riddled with boarded-up buildings and trash-strewn empty lots, and you get an ugly sense of how our civilization's balance is failing.
- We can harvest or extract biological resources in a few hours, but in some cases can't replenish them in even a few years or—in some cases—centuries. Where are the cedars of Lebanon? The chestnut and black-walnut trees of Pennsylvania? The elms of Elm Street, where my parents lived in New Jersey when I was born? Half of the forest cover that let the earth breathe at the time of Jesus is now gone. In the mid-twentieth-century America depicted by Norman Rockwell, we were given idyllic views of our natural abundance, like those *Saturday Evening Post* illustrations of life on the family farm or of a boy sitting on the bank of a pristine creek, fishing. Those idyllic memories have long since been blown away by the realities of industrial agriculture, industrial logging, industrial turkey farms keeping thousands of birds alive in squalid conditions by filling them with antibiotics, and industrial fishing operations now cleaning out the oceans. For example, the commercial fishing industry kills more than seventy million sharks per year just for their fins, used to make soup for Chinese gourmands. After the fin is cut off, the shark is thrown back into the ocean, where it drowns. A California science teacher, Judy Ki, puts this practice in perspective for the *Los Angeles Times:*

  > It's not right to slaughter massive numbers of sharks for a bowl of soup that lasts for five minutes. . . .Extinction lasts forever.

"Dragnet" was once an entertaining TV show about tough cops, but its real-life referent now is a kind of high-tech fishing—a huge ship equipped with hydraulic pumps dragging a giant net that can encircle a school of fish and haul out 400 tons of fish in a single scoop. There's no way the fish population can keep up. And then there's bottom trawling, where nets are weighted and dragged across the sea floor. Along with the fish captured for human consumption, the process hauls up bycatch—other life that has constituted the whole sea-bottom ecosystem until the moment of the catch. In a recent year, US bottom trawling dragged up 800 million pounds of sea life, of which 720 million, like the bodies of discarded sharks, was bycatch.

- We're very pleased with our abilities to multitask, but the technologies we've built give us the illusion that we can use those abilities at speeds we're not programmed for—often with fatal consequences. A study at the University of Utah, commissioned by the American Automobile Association (AAA), found that hands-free, voice-activated technologies used in cars by drivers trying to send e-mails or Facebook messages while driving are distracting enough to have caused an epidemic of crashes. AAA estimated that the number of cars in the United States that are equipped to send voice-activated messages would increase from nine million in 2013 to sixty-two million in 2018. Our survival skills are now heavily compromised by our chatter.

We might think we can solve that problem by designing driverless cars, but driverless cars are a bit like lethal weapons in the hands of people who are sometimes careless, sometimes drunk, and sometimes psychopathic. The reliability of a tech system is no better than that of the human agents who manage it—witness the fate of a mechanically flawless commuter train that flew off its tracks and killed four passengers in New York when its dozing driver took it through a sharp 30-mph-limit curve at 82 mph. A driverless car system, if built, will be far more vulnerable than trains, which usually stay on their tracks. At Senate hearings on the future of driverless technology, in 2013, questions were raised about the risks of "catastrophic cyber-attacks" on such cars. Senator John Rockefeller, of West Virginia, raised the specter of a car somewhere in the United States being hacked and crashed by "some fourteen-year-old in Indonesia." Or, perhaps, some disillusioned-with-the-world, nineteen-year-old ISIS recruit from Indiana, now in Syria?

- We can retrieve information thousands of times faster than Thomas Jefferson could, but can't comprehend it any faster at all—and probably not *as* fast as Jefferson could. After all, T.J. did a lot more reading than Americans do now. To read well (as opposed to reading phone-texts, tweets, mini-articles, quick blog comments, and TV listings), you have to take your time. Reading well, like eating well, requires slowing down. For our persistence-hunting ancestors, reading the signs left by a fleeing animal meant *not sprinting*, unless you were being attacked by a saber-toothed tiger, and by then it was probably too late. Nanny-techs that seem friendly and innocuous, but compel us to sprint in our perception and decision-making, are making us stumble in catastrophic ways.

•

I mention Thomas Jefferson not just because he was a prolific reader and had the largest personal library in America, but because *despite* the fact that no horse-drawn carriage could take him more than maybe nine miles an hour from Washington to New York, versus the 500 mph we get on an airliner, and *despite* the fact that he had to use his library to look things up whereas we have search engines many thousands of times faster, Jefferson was able to accomplish more in terms of preparation for his country's future than any of our current leaders seem able to do. Jefferson was the lead author of the Declaration of Independence and a strong influence on the framing of the Constitution (calling for limits on the power of the central government), which still serves as the country's primary guide to lawful citizenship two and a half centuries later, whereas our incessantly rushed and myopic leaders have trouble formulating coherent guides even to the next few years.

President Kennedy once invited a group of Nobel Prize winners to dinner at the White House, and toasted (or roasted) them with these words:

> I think this is the most extraordinary collection of talent, of human knowledge, that has ever been gathered together in the White House—with the possible exception of when Thomas Jefferson dined alone.

A clue to the difference between Jefferson's productivity and that of almost anyone in our sprint economy is suggested by a bit of advice he offered in a letter to his daughter, that might be applied to his reading versus our Tweet-sized communications, but also to our increasing passivity and inclination to let the nanny-techs do the work we Americans once did with our hands, backs, and brains. "Determine never to be idle," Jefferson wrote. "No person will have occasion to complain of the want of time, who never loses any. It is wonderful how much may be done, if we are always doing." He offered similar advice, regarding the time it takes to read a book, in his 1781 "Notes on the State of Virginia":

> But . . . time is not lost which is employed in providing tools for future operation.

That perspective on time—and, implicitly, on the rush of modern life that for all its speed isn't getting us any closer to a viable "future operation"—provides an insight to a paradox that may now be critical to our survival. The paradox, well understood by many athletes (and not just endurance athletes) is: *"To get there faster, slow down!"* Or, in the broader context of the human future, to get there *at all*, slow down. A few years ago, the editor of *Men's Fitness* magazine, Roy S. Johnson, wrote:

> Great athletes . . . say the game 'slows down' for them, particularly at critical moments. That's why a baseball player or tennis player can read the spin of a baseball or tennis ball when it looks like a blur to the rest of us.

To elaborate on that point, I like to draw a connection between two very different men who are both icons to millions of Americans—a Founding Father and a Superbowl quarterback. After quarterback Aaron Rodgers led the Green Bay Packers to victory in the 2011 Superbowl, the Packers coach Mike McCarthy explained his performance thus: "He is at the point in this game that the game has slowed down for him." It's not that the great athlete is literally moving more slowly, but that he has learned to slow the action around him in his perception, so that he can make more deliberate, focused, and effective moves of his own. If he doesn't waste time or energy being frantic, distracted, or inattentive, he can sometimes be amazingly effective. It's just what Jefferson said. We may be vastly faster than TJ in our travel time or our acquisition of information, but we doze or watch sappy movies while riding on a plane, and spend as much time in the "shallows" of the Internet as Jefferson spent reading for depth. And while Jefferson and his colleagues tried to look centuries ahead, our ever-surrounding nanny-tech assistance has made us so mentally and physically slack that we can't seem to envision what our needs will be even a few decades from now.

Power and Speed without Synchrony

The mismatch between our tech-assisted capability and our original nature is so large that we are all stumbling—our top politicians and corporate executives as much as anyone. If we were stumbling only from that mismatching, and still had clear vision, we might merely be like a very tired ultrarunner who's losing coordination late in the race but still envisions the finish line ahead—and will still get there. But now we are stumbling in the dark. The lack of synch between our nature and our high-speed techs is exacerbated by a lack of coordination among the techs themselves.

The result is a kind of frustration we experience almost daily. You want to do a simple task, but what once would have been accomplished by a single well designed tool, or by your working directly with another person, is now done by a series of interconnected techs designed to make the process more efficient—or, at least, more efficient at making profits for the people running the system. Too often, the result is an exasperating waste of time.

My first awareness of how the interfacing of smart technologies can produce truly maddening results came decades ago, when stores were starting to use fully computerized checkout systems for their cashiers. One day I bought an item that cost eighty-nine cents, and handed the cashier a dollar bill. At the time, I was in my early thirties, and had had long experience with this sort of half-conscious transaction: you hand the cashier a dollar, she hands back eleven cents, you say thanks, and you're out the door. On this particular occasion I was in a bit of a hurry, and automatically held out my hand for the change. Instead of giving it to me, the young woman stood staring at the cash register, chewing gum and waiting for it to tell her how much change to give me. I withdrew my hand awkwardly. The eleven cents wasn't important to me, but I felt an inculcated compulsion to be polite and do what the system expected: just wait. However, there were a couple of other customers waiting behind me and I could sense their agitation, so I asked, "Could you just give me the eleven cents?"

And then it dawned on me that maybe she couldn't, and/or didn't *know* it was eleven cents. I'd been in the habit of doing simple adding and subtracting all my life, but I'd also heard that kids in school were being allowed to use calculators even when taking math tests, and maybe doing a simple subtraction wasn't automatic with her. But at the same time, I realized that whether or not an eighteen-year-old cashier can quickly subtract eighty-nine from one hundred with confidence, the cash-register system she was now using had other purposes she couldn't override. It calculated deductions for sale items or coupons. It kept track of sales taxes paid. It made inputs to inventory. It printed receipts. It deactivated the exit door alarm for purchased items. And it opened the cash drawer for change, when those other things were done. If I wanted my eleven cents, I had to wait for the computer to open the drawer. Usually, the process was instant. But if there was a delay, there was nothing to do but cool my heels. I left the store marveling that an employee in her position might actually have needed a computer to subtract eighty-nine from one hundred, and thinking that if that was how she always *had* to do transactions, the job wasn't doing much for her future.

That was just a few minutes lost for me. But on other occasions, having to deal with an economy that involves imperfect interfaces among multiple technologies—with overseeing humans nowhere in sight—has caused me to lose far more than a few minutes. Some snafus have cost me hours, even days. Cumulatively, such occasions have taken away a lot of my life. One glitch, involving a hospital and the Social Security Administration, cost me a protracted total of more than forty hours of phone calls, emails, irritable letter-writing (plus trips to the Post Office to request proof of delivery), and just plain worrying, over a span of more than a year, to untangle—and if I hadn't been abnormally persistent, would also have stuck me for $11,000 I didn't owe and didn't have. All because of a glitch I identified at the outset, and could have resolved in five minutes had there been a live person available with the authority to see the whole situation and act on it. But all the live individuals I managed to reach had job descriptions that did not include any such authority: they'd been assigned specific tasks that involved specific bits of information that were outputs from, or inputs to, other particular tech systems that were run by other individuals with strictly compartmentalized job descriptions.

That was an ordeal, and even though the original transaction if done correctly by a person with a modicum of authority might have taken five minutes, it infected a year of my life with stress because it was under the autocratic control of a phalanx of technologies that were perpetuating an error that none of the individuals I was able to reach were responsible for. The irony is that while each of those people's jobs relied on technologies that accomplished some of their tasks hundreds of times faster than in the decades when I was growing up, when put together they sometimes took hundreds of times *longer* to get the job done right than we could have half a century ago. A lot of time I could have spent reading, thinking, talking with my family or friends, or hiking on the Pacific Crest Trail, got consumed by the frustration. Eventually the dispute found its way to someone who *did* take responsibility, and *did* resolve it in minutes.

My hospital-bill hassle was not an unusual case. Even if all the systems involved in modern life work correctly 99 percent of the time, the problem is that in a typical day the typical American relies on many hundreds of tech systems. We have *systems,* not stand-alone devices, processing our income tax returns, car registration renewals, phone calls, sewage disposal, mortgage payments, water supplies, cloud computing, power grids, power steering, car warranties, traffic lights, bank accounts, plane reservations, air traffic control, cabin pressure, traffic lights, insurance claims, and scores of others we never hear or think about. If just 1 percent of those many hundreds of systems we rely on yield glitches in a given day—well, 1 percent of many hundreds is many. The time you take to cope with them is time stolen from more meaningful pursuits, and if nothing else it's exhausting. As a trail runner, I know that when you get tired, you can stumble. A lot of us are stumbling.

It would be a colossal mistake to think the problem here is just "glitches." More pervasively, it's entire systems, or even industries, failing to mesh not only because those systems are now so vast that no agency or company can afford to closely monitor all the possible ways they can fail, but also because they're the products of a sprint economy in which it's every hustling entrepreneur or venture capitalist for himself, with the first one to get to market (or to the Patent Office) reaping the lion's share of the rewards.

We have a long history of incompatible systems: English and metric measures (and tools); broad and narrow-gauge railroad tracks; Beta and VHS videotape formats; European and American voltages and appliance plugs. Most of these incompatibilities are eventually resolved either by standardization or by establishment of separate markets (Mac and Windows, oil and latex paints, digital and film cameras). With the technological tsunami of the past few years, however, change is coming too fast for such orderly shaking-out.

For example, the established pace of change in automotive design is out of step with the much faster pace in the design or upgrading of digital devices. Here's a snippet from a report by journalist Jerry Hirsch, in the *Los Angeles Times:*

'Infotainment' Systems Out of Sync

> The latest in-dash 'infotainment' systems are turning into a giant headache for drivers. Problems with phone, entertainment, and navigation functions were the biggest source of complaints in the latest J.D. Powers & Associates survey of new car quality, easily outstripping traditional issues such as fit and finish and wind noise.

Hirsch points out that a big part of the problem is the "disconnect" between the product cycles of cars and smart phones. Buy a new car, and you can get parts and service for at least ten years. A smart phone is out of date after three. Yet, even that mismatch is fairly simple compared to others that have begun to plague American life—such as the enormous mismatches between scientific research and public education, or between social needs and government budgeting.

In civilization's incessant battle with entropy, there are always new advances toward better coordination between disparate systems—such as universal building codes, international treaties, stock exchanges, and air traffic controls. A recent advance was the collaboration of seven motor vehicle manufacturers to develop an international standard for electric vehicle fast-charging systems to sharply cut the time drivers spend recharging their batteries. Yet, significantly, the larger purpose served by this development is the demands of the sprint economy—faster growth in e-car sales and more pressure on other systems to change faster as well. Easy e-car recharging is a promising example of human cooperation winning a battle, but it appears more and more likely that entropy is winning the war.

For the Far-Seeing

Contrary to the urgings of AT&T, Intel, and Time-Warner, there's no intrinsic benefit to achieving greater speed in all things. Often, as *devices* get faster, *systems* get more backlogged or bogged down—and we bleed our brief time on this earth into the frustration of trying to cope. When we can rediscover the rhythms and cycles that connect us to the cycles of the natural world on which our lives entirely depend, paradoxically our capacity to live well and long will be optimized.

Of course, the admonition to "slow down" is one of those pithy bits of advice that prove far easier to agree with than to practice—like the

Golden Rule or the call to be "mindful." How do you slow down when everything around you is speeding up? How can you be confident you won't just fall further behind?

It's helpful to recognize, first of all, that Western cultures have trained us to think in polarities (good/evil, fast/slow), not in paradoxes. In thinking about survival, it's worth taking a little time away from the hurly-burly to study how this slowing-down paradox works. Consider the physiology of those situations where your mind is *racing*—in panic, urgency, rage. If the celebrated airliner captain Chesley ("Sully") Sullenberger had panicked on that day in 2009 when his Airbus struck a flock of birds over New York, causing both engines to fail—if his blood had been flooded with adrenaline demanding forceful, fast physical action by a man who had to stay in his seat, the risk of his jerking the controls too abruptly or hard would have spiked, or his fear of making wrong moves might have made him too hesitant to operate controls with the right touch. Instead, he remained calm and focused in a way that—as has been documented in many such cases—enabled the scene to play slowly enough in his perception to let him make the right moves with due consideration and deliberation, setting the plane down neatly on the Hudson River with no lives lost.

Recall my account of the quarterback Aaron Rodgers, handling the ball with similar calm deliberation—enabling him to thread his way through the turmoil of colliding bodies as if that colliding were all happening in slow motion. That kind of *slowing of what you perceive* can be practiced amidst the rush of everyday life, as airplane pilots and quarterbacks have done countless times in routine training. For me, a useful way to practice this discipline is to use my body as a messenger to the brain. If I'm in a rush because I'm late, I'll deliberately slow my walk, and somehow that signals to my glands and nerves to quiet down—and then, the loss of a few seconds in a slower walk lets me see more clearly (or less chaotically) how to navigate the coming minutes or hours with optimal efficiency and focus.

In the coming time of climate disruption, public panic, and failing systems, this practiced skill of slowing—and focusing—will be essential to us.

7

The Footprint of the Future

We Are Not Cows!

Cowboys epitomized the original and now romanticized American ideal of the independent, tough, don't-fence-me-in life. They were the heroes of 1950s TV westerns—the predecessors of the tough-cop, tough-FBI, tough-mafioso roles of the more urbanized America that took over. They were also the predecessors of the anti-government (or more politically euphemistic "limited-government") movements that have resisted regulation in all its intrusive forms. Ideologically, you might expect cowboy types to be the people most *antagonistic* to the nanny-tech nation that treats its citizens as ever more in need of guidance and protection. But that's not the way the economy works now. In the twenty-first century, guidance and protection—and never-ending entertainment—are what generate the profits that keep the cowboys in the saddle. The nineteenth-century cowboys were libertarian; the top guns now are more authoritarian.

It was the University of Colorado economist Kenneth Boulding who in 1966 first called the American way of business the "cowboy economy"—referring to a system characterized by minimal government regulation and profligate use of nonrenewable resources. Two years later, Buckminster Fuller used the term in his book *Operating Manual for Spaceship Earth.* As the inventor of the geodesic dome, Fuller was anything but anti-technology, and in fact was one of the more visionary techno-innovators of his time. But, like the scientists who would issue the *World Scientists' Warning* nine years after his death, Fuller was profoundly concerned about the rate at which we were burning fossil fuels. He believed that we humans can be successful with our industrial progress "provided that we are not so foolish as to continue to exhaust in a split second of astronomical history the orderly energy savings of billions of years' energy conservation aboard our Spaceship

Earth." The cowboy economy was Fuller's and Boulding's metaphor for a hell-bent ride toward ecological exhaustion.

Boulding was concerned about the ecological myopia of the country's industrialists—about the extractive- and manufacturing-industry cowboys who, so to speak, ride the range and regard it as theirs to do with what they will. Implicitly, that was a concern about the cowed creatures those industries are exploiting—about *us.* Like the cattle that were once driven by the cowboys' whips and now stand or lie in reeking feedlots awaiting their reward, we are far removed from what our wild ancestors were. The prehistoric predecessor of livestock cattle was the aurochs, a wild bovine more like a buffalo than a farm animal, and one which a cowboy would have found very difficult to herd. The ranch cows we have now have been tamed—cowed—by a hundred centuries of breeding, and are fairly easily controlled. They have lost their wildness.

Modern humans, too, are heavily controlled. Some people are as domesticated as those little dogs you see leashed on Manhattan sidewalks, whose progenitors were once wolves. Since Neolithic times, we too have left the wild far behind. With the dawning of civilization, the epic transition from the hunting and gathering of wild animals and plants to the raising of domesticated animals and crops, we also tamed our own species. To replace our hunting and gathering, we gained new skills at farming, building, and developing the specialized occupations of trading, accumulating property, establishing territories, and managing our growing populations. And as we gained new skills we also began to lose the skills of survival in the wild—the ability to track game and identify hundreds of wild plants and to escape or fight predators. We have largely forgotten that the wild world is vastly more complex than the one we have built.

Yet, complexity is the real measure of higher life, and the source of our original drive to seek, discover, and invent. Scientists tell us that life is organization; death is *dis*organization, or entropy. Our modern technologies, especially the most recent ones—the ones that take us off our feet and most relieve us of any sense of needing to think hard or observe sharply—have done us a mortal disservice by convincing us that they give us a world of vastly greater complexity than our prehistoric predecessors could have known.

What we actually live with now is gargantuan *complication,* with only here and there an exceptional work of complexity of the kind that a Mozart, Einstein, or Darwin might achieve but that most of us will never know because (1) we're too bogged down in the complications of new techs that don't mesh with each other or with the rhythms of the natural world, and (2) we are all specialists now, uninformed about the realms of the world with which our jobs don't visibly connect, or with which the speed of our work doesn't mesh. Those effects of domestication—separating us from the real complexities of the wild world, and narrowing our knowledge of it—make us ever more easily herded.

A telling indication of how much prehistoric peoples knew about the complexities of the wild world is the documented knowledge of indigenous tribes who still pursue the hunter-gatherer life—or at least did until very recently—in a few of the remaining wild places. In his book *The Tender Carnivore & the Sacred Game*, the anthropologist Paul Shepard noted, albeit in the slightly dated language of a few decades ago:

> The extent of this basic knowledge has repeatedly astonished scientists working among modern primitive people. One group of Colombian Indians knows two hundred species of a single genus of plants by sight, name, medical, nutritive, ecological, utilitarian, and symbolic use. As many as two thousand plants are recognized by many groups. It is clear, then, that the so-called savages are not intellectually impoverished as Western culture has so long believed.

A few years ago, one of my colleagues at Worldwatch suggested, with chagrin, a corollary fact: that a typical American teenager today might be familiar with hundreds of models of cell phones, laptops, apps, video games, jeans, and other commercial products, but be able to name only a few wild species by sight—and even then with very little knowledge of their ecological roles or uses. Yet, there may be more complexity in the biology and ecology of a single wild bird or plant than there is in all those hundreds of commercial techs or products combined.

Where does this kind of gross distortion in our knowledge come from? I was watching TV one night, and a new commercial came on—a bus full of children, headed out for a field trip. A man at the front of the bus, a forest ranger, holds up an oak leaf and asks the kids if they know what it is. No response. He explains what it is. Then another leaf. The kids are bored. And then, as if he was just fooling with their heads (he was), he says:

> "We're not going to the forest, guys—*we're* going to . . . *TOYS 'R US!*

The kids break into raucous cheers. The message, with an adult's (and major American corporation's) approval: *Nature sucks!* The kids are suddenly animated, and ready to affirm that life is not about things that are alive, it's about stuff you can buy.

What happens to kids like that when they get a little older? One thing, I suspect, is that they become easily manipulated by advertisers for products other than toys (though adults now spend even more on toys for themselves than for their kids). For example, there's this recent TV commercial, making what strikes me as a fairly pathetic suggestion:

> "It may seem strange, but people can love their laxative.... *Love* your laxative. MIRALAX.

Well, yes, it *does* seem strange. Can you have the same exalted feeling about something that makes you run to the toilet as you do about, say, the sight of a red-tailed hawk riding on the wind? I've always thought the idea of "love," beyond its references to close human relationships, has been used most meaningfully in references to the larger relationships we have with the mysteries of creation, or with the living world that sustains us. Edward O. Wilson, in his 1984 book *Biophilia* ("love of life") wrote that humans have an instinctive need to connect with the other life of the planet—and that our love of nature is not just aesthetic, but is a biological bond that has always been critical to our survival. So, we love our mountains and lakes and oceans. But laxative?

The prehistoric human may be stereotyped as a simple "cave man," but actually needed a voluminous knowledge of the wild world in order to survive. He or she arguably gained a far clearer view of the living world than most of us have now. "Can we face the possibility," Shepard asks, "that [prehistoric] hunters were more fully human than their descendants?" A more hopeful way of putting that is to consider the possibility that for humanity to survive in the end, we will have to completely relearn what that "fully human" love of life really is, and teach our children not to confuse the things we love with the stuff we can buy in a drug store.

For the Far-Seeing

Recognize that the complications and stresses of our lives are exacerbated and perpetuated by the political battles that dominate public debate about "issues," and that ideological abstractions are dangerously distracting us from the physical and biological catastrophe happening right now under our feet and in the air we breathe.

Rather than reflexively taking sides in the seemingly intractable divide between libertarian and liberal politics, for example, consider their respective roots in the history of civilization, and then think again about the politics of survival. Recall that the *libertarian* view has roots in the cowboy economy that prevailed in a country where the prairies and fresh air seemed endless—and where the don't-fence-me-in perspective was a natural response to encroaching population and acquisition of property. Then recall that the modern *liberal* view had roots in a period when human population was exploding and the Industrial Revolution was accelerating the millennia-long process of creating new industrial occupations, and when new dangers (of accidents, injustices, encroachments) were proliferating to the point where far more regulation and protection was demanded.

Then consider that the perceived enemy of both the libertarians and the liberals has generally been one form or another of authoritarian rule, though in different guises. The libertarians worry about overreaching regulators and bureaucracies; the liberals worry about the dictatorship of the rich and the privileged. But both are blinkered by their near-term preoccupations—and failing to acknowledge a far more pervasive and far-reaching threat to *all* of us: the demonstrated inability of any form of government we have ever known to govern sustainably.

For Americans, maybe it's time to recognize (1) that conservatives and libertarians are *not going* to cut big government down to size as promised by their rhetoric; the weight of the federal footprint is far too great, and their own power is far too dependent on the very system they ideologically oppose; and (2) populist or liberal sentiment is *not going* to equitably redistribute wealth, if doing so requires the consent of the wealthy who control the democratic process. These things are just not going to happen in the time it takes for the climate to take down far too much of what we've built, or for any of at least four other scenarios of collapse (Chapter 8) to play out.

Practical Options

- To disentangle from the mounting complications imposed by the cowboy/sprint culture, cut way back on the time you spend worrying about political issues, government overreach, or corporate exploitation—and use that time to go outdoors and reconnect with the living world you'd like to stay connected to when that is all you have.
- To reassure yourself that the failing of civilization need not be all-obliterating, note that even as big government and corporate economics are faltering, many small-scale civil society groups are succeeding. Groups

like your local Sierra Club, Doctors Without Borders, and other socially or ecologically conscious groups won't save the world, but could play a decisive role in saving *you*. They are ways to make human connections that will be critical later on. I recall President George H.W. Bush talking about "a thousand points of light" in his inaugural speech, and I wasn't sure whether that was just political rhetoric, but it's an apt image for the good being done in a thousand places by groups that have negligible power in the current hegemony but are the sparks from which new human powers may grow (see Chapter 9).

8

Seeing Past Dystopia

Facing up to Malthus, Orwell, Super-Storm Sandy . . . and What?

My wife Sharon and I were channel-surfing one night—yes, looking for mindless entertainment like almost everyone else—and stumbled on one of those serious PBS talk shows we tend to avoid when discouraged or weary. A man was talking in somber tones and his face looked familiar. Then his name popped up on the screen: Edward O. Wilson. In my work at Worldwatch, I'd had Wilson's work in my peripheral vision for years—he is one of the world's most far-seeing biologists, and one of the leading experts on biodiversity loss. Some of his research had indicated that staggering numbers of species are going extinct every year. Since we humans depend on other species for all of our food and air, there's a looming (though rarely publicized) question of just how far extinctions can go before whole ecosystems—and eventually even the biosphere—will collapse. I had also been aware that that possibility was only one of at least five major scenarios discussed by scientists that could realistically lead to the end of civilization, or at least to its mortal wounding (it could take a few years to go down).

As I watched the TV forum in a sort of stupor, the camera switched to another professor I recognized, and then another. Weary or not, I was transfixed. These were not people you often see on network TV, maybe in part because what they have to say makes the masters of the world economy, who control the bulk of funding for TV, seethe. Then, another professor was being interviewed and I didn't catch his name but his comment stuck me like a thorn. I didn't have pen and paper within reach, so I may not recall what was said exactly. But it went about like this:

> *Interviewer*: Do you think humanity has a chance?
>
> *Professor*: If you gave me a truth serum and asked me that, I'd probably say "We're cooked." But if I woke up the next day and thought about my grandchildren, I'd probably say I have to take it on faith that there's a chance, and then I'd get back to work.

It was almost exactly what I had said to myself and my family and closest friends. At least, I had used that same phrase: "We're cooked." And rationally, I have to admit, there's a lot of evidence that we are. But like that professor, I'm continuing to work on the assumption that as a mere human I can't know anything with absolute certainty, and while the pot may be unmistakably starting to bubble, I can look around me and see that we're not actually boiled yet. Well yes, a few hundred million people have been killed off prematurely (by war, crime, disease, drugs, genocide, starvation, and neglect) in just the past few decades, but billions more of us are still here wondering what our fate will be—or in too many cases not wondering enough.

While almost no-one whose income or position depends on how he or she is perceived by the general public will go on TV and say "we're cooked," people who have been privy to what's really happening have been getting bolder in print. Ellen V. Futter, president of the American Museum of Natural History, in announcing the museum's survey of the global biodiversity crisis a few years ago, wrote:

> It is imperative that we act now to bridge the gap between scientific knowledge and public understanding, not only *to ensure our own existence*, but from an inescapable moral responsibility to future generations and the planet itself. [italics added].

And the Stanford University biologist Paul Ehrlich, co-authoring with Anne H. Ehrlich, published an article in the *Proceedings of the Royal Society of London* in 2013 stating flatly that "Now, for the first time, a global collapse appears likely."

The Five Primary Nightmares: How Probable Are They?

When I sometimes let go and have a brief fling with resignation and think about the various scenarios of civilization's collapse, I find myself playing the dark-humored game of "What will get us first?" I wonder this partly because it's a question that I don't think very many of the *scientists* ask; they're too resolutely focused in their own fields to be confidently informed about others. At the Worldwatch Institute we were generalists, tracking broad global trends. And while the downside was that we knew relatively little (compared to the leading experts) about any of the phenomena we were tracking, what we could see was more panoramic.

I also find myself drawn to this exercise because all of the major dystopian scenarios have a common theme: the abuse or misuse of technology. An ocean of evidence now tells us that contrary to the teachings of most religions, it's not the evil in us or our human adversaries that is condemning us, but our often reckless attempts to expand our powers with technology. Our original powers are very durable; the tech-leveraged expansions are precarious. We are the incarnations of Icarus, that man of myth whose ambition waxed greater than his ability to envision outcomes.

Quick aside: There are other end-of-world scenarios we hear about besides the five I'll describe here. There've been those sober reports about the possibility of the earth colliding with a mountain-sized asteroid, or about a loss of our magnetic shield exposing the planet to solar flares of a magnitude even Dante could not have imagined, or about the chances of a near-earth gamma-ray explosion. There's the much talked-about speculative idea of the "Singularity"—that fateful moment when our computers surpass us in intelligence and begin to seize power. But the probabilities of these are all very small. For the following five, the probabilities of civilization-destroying outcomes are no longer small:

- **Cascading extinctions:** As is now well documented, Earth is presently in the throes of its sixth mass-extinction event—the first since the dinosaurs were extinguished sixty-five million years ago, and the first to be caused largely by human activity. In 2011, geologists at Brown University and the University of Wisconsin confirmed what other scientists had long inferred or suspected: that in at least two of the planet's five previous mass-extinction events, the loss of global biological diversity triggered collapses of whole ecosystems. If that should happen again, on a global scale, our food supply could be mostly wiped out, along with much of our capacity to resist decimating epidemics.
- **Nuclear holocaust:** At the peak of the Cold War, only the United States, Soviet Union, and United Kingdom had nuclear bombs. Now, so do China, France, India, Israel, terrorist-infiltrated Pakistan, and rogue-state North Korea, with Iran a dark-horse candidate to join the club

whether welcomed or not. We reached the brink of nuclear war during the Cuban missile crisis of 1961, after which thousands of Americans dug underground bomb shelters in their backyards—in retrospect, a little like a SWAT cop putting on an extra sweat shirt to stop bullets. I can still remember the teacher at my elementary school in New Jersey, a half-hour train ride from New York, telling us kids that if the A-bomb siren sounded, we should get under our desks. She was kind enough not to mention that if an atomic bomb actually fell, it would vaporize both us and our desks, not to mention our school and town. Fears have receded since then, but only because public officials and media have directed them elsewhere. To the nuclear scientists, the danger is greater now than when President Kennedy faced off with Nikita ("We will bury you") Khrushchev. In 1960, the "Doomsday Clock" of the Union of Concerned Scientists—a symbolic clock face used by the *Bulletin of the Atomic Scientists* to gauge "the state of existential threats to civilization"—was set at seven minutes to midnight. In 2015, it was adjusted to *three* minutes to midnight, the closest we have ever come to annihilation, in the consensus opinion of those who should know. There are a number of conceivable events that could move the clock its remaining three symbolic minutes in a single real-time hour.

- **Global Pandemic**: The flu epidemic of 1918 killed fifty million people—at a time when international travel and trade, the conditions that let viruses spread around the globe, were only a tiny fraction of what they are now. The advent of flu vaccines and antibiotics assuaged public fears of a global pandemic occurring in the future, though movies like "Outbreak" and "Contagion" keep us titillated with the possibility. According to Jason Tetro, head of the Emerging Pathogens Research Centre at the University of Ottawa, "the world is increasingly likely to see a major event." Tetro explained, for the website offgridsurvival.com:

> It starts with migration of agriculture and urban environments into more rural and remote areas, increasing the likelihood that a potential pandemic strain of a pathogen will come into contact with humans.... Then, thanks to the rise in densification of both animal and human populations, these pathogens can spread.... Finally, with travel from the localized area, the pathogen can move worldwide. This fact is particularly important when one thinks that almost one hundred years ago, when we had the 1918 pandemic, it could take months to circumnavigate the globe. Today, it can be accomplished in a day.

That possibility graphically illustrates how our technologies have gotten out of synch with the rhythms of nature. Microorganisms can reproduce—and make evolutionary adaptations—faster than humans, even in the fever of the sprint economy, have been able to develop new antibiotics.

The result is that many diseases are resistant to antibiotics, and a few are virtually immune. The World Health Organization warned in 2012 that untreatable strains of gonorrhea are spreading across the world. And Sally Davies, Britain's chief medical officer, confirmed in 2013 that, as reported by Kate Kelland in *Huffington Post*, "Cases of totally drug-resistant tuberculosis have appeared in recent years and a new wave of 'super superbugs' with a mutation called NDM 1, which first erupted in India, has now turned up all over the world." In 2014, the most feared of all contagions known so far—ebola—gave the world an unnerving reminder that epidemics are not just the stuff of disaster movies.

Smart as we humans may be, we're slow-footed (recall Chapter 6). Viruses and germs are nimble. New diseases have been emerging with growing frequency, and if a particularly virulent one that is immune to our drugs should get out of control, the result could dwarf the flu of 1918. (Global population now is about six times what it was in 1918, which would make an equivalent death toll more than 300 million—but as noted, that's before we count the huge added vectors of millions of people traveling around the world every hour, carrying their germs and viruses.)

- **Malthusian population collapse:** Thomas Malthus published his *Essay on the Principle of Population* in 1798, warning that unchecked human population growth would ultimately lead to population implosion. A quick look at his argument, aided by a couple of centuries of hindsight, yields this: population grows exponentially, but the means of sustenance on a planet of finite size cannot grow much at all, and ultimately will more likely shrink.

At the time of Malthus's essay, the world population was under 1 billion—meaning that over a period of 100,000 years before then, human population had increased by an average of less than 10,000 per year. In a little over two centuries since the essay, population has ballooned by more than six billion, or an average of roughly thirty *million* per year—*three thousand times as fast* as it did over the thousand centuries before Malthus had his insight. Remember Han Solo, in the first "Star Wars" movie, shifting his spaceship to hyperspeed? Human population expansion has gone into hyperspeed, and if not interrupted will continue until the mid-twenty-first century. And meanwhile, the food-producing capacity of the earth has expanded only a little. And that gain is largely nullified by the fact that the amount of arable land per person has shrunk.

Over the past few decades, the technologies of nitrogen production and the Green Revolution (fertilizers, plant-breeding, genetically modified foods, etc.) have increased crop yields and bought us a little more time, but meanwhile the technologies of advertising and consumption have used up that time and more. What the original agricultural revolution of early Biblical times hath given us, Burger King and Coca Cola hath taken away. Malthus envisioned collapses in the forms of resource wars, famine, and disease. In the twentieth century, we saw the harbingers: the epidemic of 1918, two world wars, the famines in East Africa and South Asia, and the now more than marginal existence of hunger even in the United States. By the second decade of the twenty-first century, about fifty million Americans were "food insecure," according to US Department of Agriculture data. A study by Feeding America, the country's largest hunger charity, found that seventeen million American children are going hungry. Even with the food-growing and distribution system still operating, the national safety net is full, or at least regarded as full by callous politics. It won't take much more burden to make it break.

To put this in an even bigger-picture perspective, and indulge in a bit of black humor that's black-hole-deep, I can muse on the following comment made by Philip M. Hauser, a commencement speaker at the Albert Einstein College of Medicine in New York, back when the global population was about half what it is now:

> If the world's population were to maintain its present growth rate for 6,000 years (about the period of recorded history), it would end in a mass of human flesh the periphery of which would be expanding at the speed of light.

Of course, as we now know, our actual expansion couldn't continue at its present rate for even 1 percent of that many years without imploding.

- **Climate Upheaval**: in my view, this is the big one, because it encompasses all of the others (severe climate events could shatter nuclear security)—and because it is already well underway. The earth's sixth extinction event has not yet destroyed whole regional ecosystems, and we have yet to see how that will play out. Nuclear holocaust may or may not be loosed upon us by the anarchy of the other scenarios, and as I write in 2015, there's still a small (but shrinking) window of opportunity to disarm the bombs before the likes of Al Qaida or ISIS get to them. A global pandemic, likewise, may or may not happen. And Malthusian outcomes will be largely determined by what happens with the climate. But climate is already in a state of alarming disruption, and its effects are cascading. When the IPCC issued its first Climate Assessment in 1990, the warning was called "bad science" by public officials who apparently feared losing their elections more than they feared losing their planet. With the release of the *World Scientists' Warning*, the politicians learned a lesson from the media: if scientists issue an alarm big enough to truly upset the dominant economy, act publicly as if that alarm never happened.

So, for a quarter of a century, America has fiddled while the world has burned. Many of the men and women who have wielded enough political muscle to paralyze US efforts to stop global warming are people who have had little or no scientific education. In its 95-0 rejection of the Kyoto Climate Treaty, the US Senate conveyed not just doubt but contempt for the climate scientists. Later, when the Oklahoma Senator James Inhofe dismissed global warming as a "liberal hoax," liberal Senators who'd voted to reject responsible leadership bit their tongues and averted their eyes. The scientists had warned emphatically that global warming would bring more intense extreme-weather events, but when Inhofe's home state was staggered by weather disasters of unprecedented intensity during the years after his deathless comment, he showed no sign of coming to his senses, apparently still sure it was all a hoax.

Inhofe's obtuseness is not particularly worse than that of scores of his colleagues, even if his comments will have caused more eyeball rolling by future historians—assuming there will be some historians among the planet's survivors. In 2014, the House of Representatives passed a bill prohibiting the Department of Defense from using any of its funding for defense against climate change. It specifically forbade any US response to the warnings of the IPCC scientists who'd been warning that climate disasters are becoming "threat multipliers"—endangering national security on multiple fronts. It was as if the Los Angeles County Board of Supervisors had forbidden the Sheriff's Department to spend any money on combating gun violence.

In November, 2013, the IPCC's Fifth Assessment revealed that the climate is destabilizing faster than the previous Assessments had projected. It warned that accelerating changes in climate will likely worsen the already ongoing human tragedies of extreme-weather disasters, war, starvation, and disease that have plagued the world during this latest 0.002 percent of the human experiment. It also warned that climate change will further widen the already destabilizing rich-poor divide. An inevitable result, sooner or later, will be massive exoduses of climate-disaster refugees in search of food, fresh water, medications, and shelter—and missing family members and neighbors. But those exoduses won't just be in far-off, phase-2-disconnected places like Liberia or Bangladesh. Consider the prospect of hundreds of thousands, perhaps millions, of people fleeing from New York City or Miami (or New Orleans *again*)—but now with no prospect of returning home in a few months. Where can they go? As if adjoining states won't already be under backbreaking strain of their own? When New Yorkers flee the rising and oft-enraged Atlantic, so will be people fleeing from cities along the entire Atlantic and Gulf coasts.

The warming will also drive migrations of *agricultural zones and ecosystems*—and with them, mosquitoes, fire ants, and invasive agricultural pests or weeds—to higher elevations or latitudes, probably leaving whole regions of farmland barren. By the late twentieth century, that process was already at work on the fast-expanding Sahara Desert and in parts of China and the American plains states and West. Sooner or later, these changes are likely to bring chaos far more sweeping and life-altering than we experienced in World Wars I and II.

As we raced into the twenty-first century, the main public response to these prospects seems to have been little more than a huge boost to the businesses of disaster movies and dystopian TV epics. On the news, even as catastrophic floods, tornados, wildfires, and other extreme-weather events became more frequent and intense, it wasn't until 2013 that the weather reporters even *mentioned* climate change. The oil and gas cowboys were still very much in control. It wouldn't be until after Super-storm Sandy and the smashing of Moore that I heard the words "climate change" finally spoken on mainstream TV in connection with the unprecedented weather events.

One quick aside, about those proliferating disaster movies: I wondered if they reflect the degree to which people instinctively sense that civilization might somehow be in trouble, even if they aren't clear about what that trouble might be. At Wikipedia, I found a surprisingly long list of dystopian movies, divided by category. I was curious about the ones which seemed inspired by real science, and which perhaps articulated fears that politicians and reporters could not. Putting aside any cyber-punk stories or sci-fi tales about alien invasions, I counted all the movies in two categories that I felt may reflect credible apprehensions: *"Post-apocalyptic"* (aftermaths of disaster, such as nuclear holocaust, war, or plague), and *"Government/Social"* (government or society attempting to exert control over free thought, authority, or freedom of information, etc.) I found several striking patterns. First, the overall number of dystopian movies has increased sharply in successive twenty-year spans, from **six** before 1961 to **thirty-five** (1961–80) to **fifty-one** (1981–2000) to **ninety-three** and climbing (in just the first fourteen years of the new millennium). Second, while there were slightly more post-apocalyptic than government/social dystopias up to 2000, that emphasis reversed sharply after 2000, with **seventy** movies in the government/social list versus **twenty-three** post-apocalyptic. Recalling the observation I made in chapter 3, about the fact that we fear people more than technology, I wondered if this inclination may have heightened apprehensions (or curiosity) about political and cultural adversaries after 2000—or perhaps after September 11, 2001—as the technologies being used to sway belief and behavior became more invasive and pervasive in recent years. In effect, fears of physical apocalypse may have been overshadowed by the fears of evil intent.

The Day I Cried Wolf

During the first years of this century, there seemed to be perilously few people in positions of influence who recognized that we were in a very dangerous situation. One was Gregory Foster, the George C. Marshall Professor of Environmental Security at the National Defense University in Washington, DC. Foster had read some of the commentaries I'd written for *World Watch*, and in 2000 he asked me to give a seminar for about 20 of our country's military analysts and policymakers. Foster headed a Pentagon group called the Defense Environmental Forum, whose mission was to persuade national security experts that environmental catastrophes now posed as much danger to the nation's security as the possibility of military or terrorist attacks. It was beginning to dawn on some of the Pentagon strategists that national defense needed to include capability to respond not only to hostile humans, but to the growing threats of crop-destroying drought, city-destroying storm surges, bioinvasions (the spread of species to ecosystems that had no evolved defenses against them), and pandemics. "Defense" was also coming to be seen, increasingly, as an ability to respond to the civil disorder that can erupt when environmental catastrophes occur.

On the day of the seminar, held at Washington's Fort McNair, we gathered in a room with a long conference table and a few extra chairs—a mix of uniformed senior officers and civilians. I don't recall everything we discussed, but one moment sticks with me. I'd been talking about climate change, and wondering if I was coming across as a credible speaker or more like an obsessive Senator Joe McCarthy warning about Communists hiding in our closets. Seeking eye contact and hoping for connection, I said, "Suppose a hurricane-driven storm surge were to hit the city of New Orleans."

Of course, climate scientists, oceanographers, and Mississippi Delta ecologists had been worrying about New Orleans for years, but mostly in academic papers that the media mostly ignored. Imagining what might befall New Orleans didn't originate with me. But here was a chance to bring the mounting risk to New Orleans to the attention of influential people who were maybe too focused on military threats to fully recognize the new threats arising. I braced myself for their response, and was soon relieved to find none of them reacting the way Senator Inhofe had. They seemed intensely interested, although I also sensed that in the slow-changing culture of the military, their ability to take action in response to new information must be heavily constrained. And they might have been thinking, "We already have the Army Corps of Engineers on the job down there, don't we?"

Back at my Worldwatch office, I sat watching a mug of coffee grow cold on my desk, wondering whether a seminar like that made any more difference than shaking a fist at a tornado. I recalled a conversation I'd once had with my colleague Michael Renner, who would later become a senior advisor to the European Institute for Environmental Security, in which we speculated that it might take a stupendous environmental disaster—like a storm-surge destroying a major city—to finally awaken the public to the reality of the climate threat. Now I dared to think it would be far better if the military, with its enormous resources, would take preemptive action—but *not* the kind of short-sighted action being taken by the Corps of Engineers, which was busy dredging and straightening rivers and building sea walls that bought time for cities but undermined the long-term ecological stability on which those rivers and cities depended.

The following year, 2001, I was invited back to do the seminar for another group. Then on September 11, a few weeks before the seminar date, the terrorists struck. Every one of my generation could recall exactly where they'd been when they first heard that President Kennedy had been assassinated. Now, a new generation, along with mine, would always recall where they were that morning of 9/11. I was in my Worldwatch office, still finishing my morning coffee, when I heard an unusual commotion in the conference room across the corridor. I peered out; the conference room TV was on and staff members were gathered tensely around it. A news camera was pointed at the World Trade Center, from which a plume of smoke billowed. As we watched, the second airplane struck.

I knew then that by the day of my second seminar, the atmosphere at Fort McNair would be forever changed. Foster suggested that I not drive my own car to the seminar this time, and instead he sent a car to my Worldwatch office to pick me up. As we entered the gate at the base, we had to drive at about 3 mph through a serpentine route lined with concrete jersey barriers—positioned to prevent any vehicle from crashing through at high speed. We were just a couple of stone's throws across the Potomac from the Pentagon, where the huge dark hole where the 757 had crashed through still gaped. Although my driver was a uniformed Army guy, we had to halt while bomb detectors probed under the car.

In the seminar, once again, I brought up my New Orleans scenario. But this time I sensed a restlessness in the group. Years later, I would wonder if the greatest damage done by the 9/11 attack, even beyond the enormous death and destruction and heartache wreaked that day, might have been the derailing of any impending US action on climate change, the impacts of which I knew would eventually dwarf what had happened at the World Trade Center and Pentagon, and over an empty field in Pennsylvania.

Four years later, Hurricane Katrina smashed into New Orleans. It caused terrible destruction, but fortunately wasn't a head-on hit and by the time it came ashore had weakened from a category-5 to a 3. In the following months, I found myself talking with Michael Renner again—about how the wakeup seemed not to have happened. Katrina dominated the news, but in mainstream coverage the climate-change connection went virtually unmentioned. What if it had happened to a much larger city, like New York or Miami, not with a glancing blow but with a direct hit? Would *that* have accomplished the wakeup?

Over the ensuing several years, a cascade of warming-driven events hit the news: the North Pole melting, South Pole shedding a piece of melting ice the size of Rhode Island, earth-scorching forest fires of unprecedented ferocity from California to Colorado to Georgia . . . and in 2012, Super-storm Sandy. At that point, it seemed, the most immediate of the five global threats I've listed was unquestionably climate. In a decade, we'd gone from half of the American population denying that global warming was even happening to at least a large plurality recognizing (despite the opposite trend among conservatives) that it was not only very real but too far advanced now to stop.

So, the most likely answer to the question of which destruction will come first now seemed clear: whatever the odds of collapsing ecosystems, pandemics, nuclear attack, or famine, we would very soon see more Katrinas—but, sooner or later, in coastal cities far larger than New Orleans. Calcutta? Guanzhou? Miami? Mumbai? Or New York again, head-on by a full category-5? Super-storm Sandy gave us a sort of premonition. A few months later, a monster typhoon in the Philippines gave us another. How many would we need?

Then again, I was also aware that the "what-comes-first" picture could change overnight: imagine a dissident Russian or Chechen group that is bitterly anti-Kremlin and has cells in the Russian military, just as anti-Washington cells are known to exist in the US military. Suppose this group, made up of angry men no older or more circumspect than the Boston Marathon bombers, obtains a black-market nuke of conspicuously Russian fingerprint and detonates it in New York with the intention of provoking a massive US retaliation against Moscow. Or, similarly, imagine an American anti-government cell in a US nuclear submarine within striking distance of the Kremlin, that would like to see the population of Washington, DC and the rest of the elitist eastern US dispatched to Hell. Or, on a different front, imagine an ebola-like virus that comes from nowhere without warning and within days sweeps the world. There's no way of knowing for sure which shock to civilization is most likely to come next, whether it be triggered by an overnight category-5 obliteration of Shanghai or Mumbai or Dhaka, or a 10-second obliteration of Moscow or New York, or the sudden outbreak of an unknown disease that kills uncountable millions. Most likely, though, is that of the five scenarios I've sketched above, the prospect of a destabilized and rampaging climate looms largest.

A Sixth Scenario

Whatever the relative probabilities of the five separate threats, which are unknowable but substantial, it seems to me now that an even greater threat than any one of those five alone—and the thing that will bring us down sooner or later, *regardless* of what precipitating shock hits first—is a Trojan Horse-like sixth: the effect of nanny-tech culture in sedating and blinding us to the looming specters of the other five. What's upon us now, aside from any looming storm cloud or potential mushroom cloud, is what the critic Thomas Mallon describes as "an app-happy Cloud of anesthetized convenience."

If a majority of the world's people were broadly educated and well informed and both physically and mentally healthy and fit, none of the lethal five threats would loom for long, at least not to a doomsday degree. As of this writing, there's still time for us to regain control of our senses. "Well educated" doesn't mean well prepared for a "tech" job, which seems to be the focus of the burgeoning industry of for-profit universities. It means having big-picture knowledge of human history, science, religion, culture, and persuasive ideas—and of the techniques used to *make* ideas persuasive. It also means learning critical long-term survival skills: hard questioning of established authorities and dogmas; lifelong seeking of better answers and understanding; insistence on thinking, not just believing what we're told to believe; and, of course, having all these things in mind as we bring up our kids and grandkids.

Most dauntingly, but maybe most important, regaining control of our senses means having the willingness and humility to recognize how much there is that we don't know. More than half a century ago, President Kennedy was giving a speech at Rice University when he made this comment, which has proved tragically prescient, especially with regard to the nation he led:

> The greater our knowledge increases, the more our ignorance unfolds.

Human nature, if allowed to thrive, encourages all of these aspects of education and discovery—and invention. Our forebears were prolific explorers of the earth for thousands of years before we had the great arsenals of technology we now have. But when our nature is relieved of its responsibilities by technologies that first assist and assuage and anesthetize it and then take over for it, it loses its alertness, curiosity, and adaptability. If that happens to large numbers of us, there's no united defense against any of these dystopian outcomes. Here are a few of the phenomena already eroding that could-be defense:

- **Loss of number sense**:

  In a tech-dependent world where human powers have been multiplied immensely, having good number sense, or sense of proportion—a particularly important form of *common* sense—is critical for the people who use these powers or may be affected by them. It will be critical for those who hope to move beyond the coming collapse. Just as surely as a grasp of numbers is a matter of life and death in the engineering of airplanes or half-mile-high skyscrapers, it's equally a matter of life and death in monitoring the rate of sea-level rise, or declines in food-producing land per capita. And in a sprint economy, it's especially critical to debunking get-rich-quick myths that have infected public attitudes about the requirements for long-term investment and survival.

  A few decades ago, kids learning math had to learn to calculate. If they now are allowed to use calculators or math apps in their mobile devices, their ability to grasp such concepts as "order of magnitude" or "exponential growth" are dangerously dulled. Anyone who isn't alarmed by the fact that the average increase in the global population per year in the two centuries after Malthus died was more than two thousand times the average increase per year over the previous thousand centuries, yet doesn't rank population in among the top "issues" humanity faces, might as well be sleepwalking across a highway. Similarly, any adult American who doesn't quite grasp the difference between a million and a billion—a common confusion, apparently—is ill equipped to participate in a democracy where citizens decide how their tax money should be spent. Apparently, to many Americans, the difference of value between a million and a billion is something like the difference in value between a Honda Civic and a Bentley—the latter just being worth "a lot more" than the former. In much of the public consciousness, that "*thousand times*" as much seems to be missing. What else explains the apparent lack of public outrage, in a country that's struggling economically, about the number of billionaires we have, each one of them having at least as much wealth as 30,000 average Americans? Or in a world where (as the nonprofit Oxfam revealed in 2014) the richest eighty-five people each have as much wealth as *forty-one million* other people? If you don't use decimals, you forget (if you ever learned) what they mean. The exuberant, uncritical network news coverage of a $636 million lottery jackpot in December, 2013 was a symptom of massive gullibility in America's understanding of what constitutes a good investment, considering how much organic produce, or good music, or repairs of deteriorating bridges or water mains, the money thrown away for that month's lottery tickets could have bought.

This isn't an argument for cutting back on calculators or computers, but it's certainly a suggestion that those techs shouldn't be used to shortcut education about what numbers and magnitudes—and *probabilities*—mean. There's a shortcut used by a majority of Americans to see what's important: just turn on the TV news. That takes much less effort than, say, reading books or research reports, or *The Atlantic* or *Mother Jones*, or the journal *Nature*, or listening closely to Thom Hartmann or Democracy Now on alternative radio. But it might also give people the very mistaken impression that murders pose a greater risk than infections caught in hospitals, or that terrorists kill more people than cigarettes, or that the American Dream is now anything more than a delusion.

- **Loss of cause-effect sense:**

  As the assault on our senses grows by the hour, it gets ever more difficult to track sequences, correlations, and connections. A young toddler staring at the fast-moving, non-human, non-animal cartoon figures (cars with faces?) on Disney TV has no idea what causes that visual blitz, which his single mom, struggling to clean house while starting a home-based Internet business in the absence of the kid's long-gone father, has turned on to sedate him. In a few weeks, he'll see that he can cause the blitz himself by pressing a button on the remote. But unless he goes to the University of Phoenix and learns to be a TV or IT technician, his grasp of cause-and-effect won't reach much farther than that button. And even if he does go to Tech U and then gets a job as a twenty-two-year-old fixing electronic equipment (if indeed there are still jobs for that by then), he'll have no idea what causes an airplane to rise into the air, or what has caused a much-larger storm surge than his grandparents ever heard of to wipe out a city of ten million.

Already, in the second decade of the twenty-first century, we're in a world where most people, including me, have no idea what causes most of the things that happen around us, because what's happening is too much and too fast and because too many of us have never studied scientific method, or because we've been told by religious zealots (or by billionaire investors who profit from our ignorance) to *distrust* science. And for the few who do care about cause-and-effect and sense instinctively that it is critical to our preparation for survival, achieving expertise in any one field now requires going so narrowly deep that there's no longer any time for breadth. Even most scientists have very little knowledge of specific cause-and-effect outside of their own fields.

- **Loss of future sense**

To have a vivid consciousness of what's likely in the future depends very much on having vivid memory of what happened under comparable conditions in the past. That's the basis of empirical science. By the same rule, knowing how to change course and alter probable outcomes depends on knowing what kinds of actions changed courses or altered outcomes in similar situations in the past. If education doesn't include history, it guts the student's preparedness for the future. If the student gets a tech education with no courses in biological evolution or the history of civilization, he's deprived of any sense of the ways in which historic momentum blows into the times to come.

•

Yet, in the face of all the foregoing, and with the slow waiting-to-exhale syntax of a sentence I once read in Melville's "Moby Dick," about the ominous "whiteness of the whale," I offer the following anti-tweet:

Whereas humanity has destroyed more than half of the world's oxygen-producing and climate-stabilizing forest in just the latest quarter of 1 percent of our time on earth so far, and is putting more time and

> money into further decimation than into stemming the ravages of the planet's sixth great extinction event . . .
>
> and *whereas* we Americans spend more of our money and time producing and watching TV shows about serial murders, sex crimes, vampires, and other forms of fictional violence than on reducing the chances of nuclear holocaust and other *real* threats to our existence . . .
>
> and *whereas* we in the industrial world spend more on headache pills, erectile dysfunction pills, skin creams, mood stabilizers, and other remedies for discomfort or anxiety than we spend on eradicating malaria and other curable diseases that kill millions, or on reducing the risks of global pandemic . . .
>
> and *whereas* the last century has been the most violent in human history, and hundreds of millions of our kind have died in this brief span via war, genocide, and other human-caused events that increase the risks of our own species being decimated by that expanding sixth extinction . . .
>
> and *whereas* our collective addictions and distractions and denials have allowed us to upset the natural life-support systems of the planet itself, via the mounting disruptions we euphemistically call "climate change" . . . ,
>
> ***Nonetheless,*** there is something in our nature that defies disaster and decimation and can do so again, now, in this ultimate challenge to our better angels.

Large numbers of people have ridden out the storm-surges of nearly all previous threats to civilization. Far-seeing individuals and groups have found ways to beat back the scourges of genocide, slavery, infectious diseases, infant mortality, air pollution, and the thinning of the ozone shield. Nations have sometimes followed but rarely led, and now again it's up to individuals and small groups or communities to lead.

For the Far-Seeing

Begin preparing, now, for impending changes that neither our political leaders nor our mainstream media are addressing or planning for, or even seem to foresee. Start with the basics: emergency preparedness for sudden, short-term disruptions, but with due consideration for the longer term.

Those longer-term considerations will come most clearly to mind, given the urgency, if—paradoxically, as discussed in chapter six—you think them through with slow deliberation. If your emergency supplies give you food and fresh water for a few weeks without refrigeration or cooking, what options do you have for a few months? Later, you'll need to think about years, and if all goes well, decades.

Right from the get-go, an extended family or community will have resources that a single individual most likely will not. So, sort through your lineup of family members, neighbors, friends, and colleagues. Who among them would be open to candid discussion of the outlook so many of us have avoided discussing? Not everyone is. Denial is still pervasive. But chances are, you know a few people you can talk with candidly, and who may have essential skills or resources. Who has a farm or a vegetable garden, or fruit trees? Who has a well? Does anyone on your list have freestanding (not grid-connected) solar or wind power?

For each of the five scenarios this chapter describes, begin to sketch out (preferably with others you trust) your options:

- If you see signs that ecological collapse is getting too close, assess the biological and ecological complexity of the region where you live, or where you might go for retirement or retreat. More biodiversity confers more eco-stability, probably even in ways we haven't net fathomed. Better to live near the rainforests of the Pacific Northwest than in those desolate parts of Kansas or Nebraska where the native grassland has been wiped out by cattle grazing or industrial monoculture (though there are counter-considerations as explained in the following chapters). Better to live near wild marine or estuary food resources (but not too close to coastal sea level!) than in a place where most biodiversity is gone.
- If you're in a place you think might be a prime nuclear target, either for missiles from hostile nations or (perhaps more likely) for a domestic-terrorist attack on a nuclear power plant or a facility like the Oak Ridge facility that stores the materials from which nuclear weapons are built, at least consider what your escape route would be if you had any warning, and where you could go to for temporary shelter. Nuclear holocaust is the hardest scenario to prepare for (if the attack is by missile, there might be no warning at all), but even marginal planning—and keeping an eye on tensions between nuclear powers or would-be powers (Pakistan vs. India, Israel vs. Iran, ISIS vs. the United States)—could spell the difference between life and death for a small number of survivors.
- With global pandemic, as with nuclear tensions, keep abreast of the news, via PBS, NPR, Democracy Now!, *The Guardian*, Wikileaks, or the leading medical journals. The biggest defenses, other than medications with awful side-effects, are separation from crowds, avoidance of travel, and a strong immune system. It's a brutal fact of evolution that disease kills the weak before the strong (more on this in Chapter 11).
- The same defenses apply in the event of more widespread deterioration of human life-support systems—such as medical insurance companies

withdrawing coverage or raising prices or going out of business. Good nutrition, cardiovascular exercise, and positive attitude are better defenses against disease or death than all the drugs advertised in TV commercials combined. Ask your doctor!

- With climate disruption, it's not a matter of if or when, but of where and how severe. Here, the ramifications for practical action are so manifold and far-reaching that I have devoted most of the last chapter primarily to them. Everyone will be affected. (But don't succumb to the temptations of the sprint culture and skip straight to that chapter, as there's more to consider first.)

For the far-seeing, this isn't a time for gloom but for girding. (If you have loins, biological evolution equipped you to help the human race go on!) On the other hand, evolution alone is no longer what keeps us alive and on course for long-run survival. As the next chapters discuss, we're in an epochal, uncharted, transition now.

9

Reigniting Imagination

In Search of Our Lost Abilities to Anticipate, Envision, and Plan

Specialists are rarely visionaries. Their vision is necessarily narrow. And the deeper a person has to go into a field to become expert, the less time and energy he or she can give to things outside the field. With every passing year, if you're a competent specialist, you have to go deeper to stay on the cutting edge. Here's a sometimes disconcerting result: an expert you depend on for an important aspect of your life turns out to be woefully misinformed about some other, closely related, aspect.

One morning I went out for a run, intending to go ten miles, as I often do—and couldn't go ten feet. There'd been nothing wrong the day before, when I'd run an easy couple of hours with no soreness afterward. But now, suddenly, my left foot was crippled. As an experienced runner, I knew my body well enough to be fairly sure this wasn't a broken bone, sprain, or pulled muscle. It didn't fit the symptoms of plantar fasciitis or any other injury I'd ever heard of. I went to my doctor, a well-respected family physician. He examined the foot, ruled out diabetic nerve damage, determined that it wasn't a deep vein thrombosis or circulatory problem, and agreed that it wasn't a pulled muscle or tendon. He took an x-ray, found nothing abnormal, and finally sat me down in his office to give me his evaluation.

"Your foot is tired," he said.

He didn't try to explain why only one foot would be tired, when the other could run ten miles with no difficulty.

I nodded mutely, went home, and made an appointment with a prominent sports medicine podiatrist. On the appointed day, the podiatrist examined the foot in the same ways my GP had, made another x-ray, and asked me into his office. He showed me the x-ray image and pointed to a very small bone. "See this? It's called the cuboid bone. There's a long tendon that comes down the outer side of your ankle, wraps around the cuboid, and crosses over to the other side of the foot. . . ." He went on for several minutes, describing in anatomical detail what was happening when my foot hit the ground—a complex business involving that long tendon from the ankle, another small bone, and excessive pronation (lateral roll) in my foot-plant jolting the cuboid out of position. "I think the pain and numbness you feel are secondary impacts on the nerves, not nerve damage," he said. He then prescribed custom orthotics to stabilize my pronation. The pain soon receded and within weeks I was back to normal running.

The family doctor had seen nothing wrong on the x-ray. There was no fracture. He had missed the diagnosis not because of incompetence or carelessness, but because despite the fact that he'd been a doctor working with pain and injuries for many years, he didn't specialize in feet. Or, more specifically, in the *movement* of feet. Among the doctors we have today, there are specialists in hands, eyes, hearts, skin, bones, blood, brains, knees and hips, feet, urinary tracts, nerves, and so on. No single doctor can be an expert on the whole body. And if that is true, then it's obvious that no doctor is likely to be very knowledgeable about, say, the health-related aspects of organic chemistry (even though the body is made up of organic-chemical compounds), or electrical engineering (even though the brain governs by means of electrochemical signals), or nutrition (even though "we are what we eat"). Or biomechanics or epidemiology or public health.

The Blindness of Experts

The same kind of occupational disconnect happens between experts in all sorts of closely related fields—between anthropologists and paleoanthropologists; between chefs and nutritionists; between aeronautical engineers and airplane pilots—or, down the chain, between airplane pilots and transportation-safety officials, or between safety officials and accident-litigation attorneys. What happens when people in seemingly *unrelated* occupations cross paths? Sometimes the effect is puzzlement or unease, but my impression is that more typically it's unquestioning acceptance. The lawyer boarding a plane bound for Washington, DC may be an FAA expert on the laws of air traffic control, but have very little knowledge of the laws of physics that will keep his plane in the air. Even if he learns that the woman in the adjoining seat happens to be a physicist, he'd be unlikely to ask her what *does* keep this plane aloft, knowing he probably wouldn't really understand her answer. And in the unlikely event that he did ask, she might reply diffidently, "Well, I wish I could tell you, but I'm not an aeronautical engineer." Meanwhile, the school teacher who tells your kid about the three branches of government in America, and the separation of their powers, may be fairly clueless about the powers wielded by corporate executives, investors, and lobbyists—most of whom, in turn, have not the slightest notion of what it's like to be a plein-air painter, backhoe operator, forensic geologist, or historian.

In short, wide vision is getting rarer. As search engines speed up research and crunching power speeds up analysis of data, and universities offer courses and degrees in ever narrower subjects, the ideal of a "well-rounded" or "liberal arts" education has become quaint and allegedly useless at qualifying a student for a job that pays. For many, the idea of majoring in English, or art, or philosophy, or history, is now a joke. A study released by the American Academy of Arts and Sciences in 2013 found that more and more students are abandoning broad education in favor of lucrative business and tech fields, and the report includes this observation:

> At the very moment when China and some European nations are seeking to replicate our model of broad education in the humanities, social sciences, and natural sciences as a stimulus to invention, the United States is instead narrowing our focus and abandoning our sense of what education has been and should continue to be—our sense of what makes America great.

After civilization began, the movement toward increasing specialization was very gradual. Prehistoric humans had been generalists—every adult a highly skilled hunter-gatherer. The agricultural revolution began the division of labor among farmers, herders, stone cutters, weavers, sawyers, builders, smiths, cooks, or guards. But new inventions such as the gun, printing press, and vaulted ceiling came relatively slowly and it was possible for very bright individuals to gain broad knowledge. As recently as the sixteenth through eighteenth centuries, at least a few people could acquire broad knowledge of history, philosophy, languages, art, architecture, religion, and science. We can marvel at the "Renaissance man" that Shakespeare or Leonardo da Vinci was . . . or, later, that Thomas Jefferson or Ben Franklin was. Significantly, those broadly educated men were also visionaries.

Maybe the last full-fledged visionaries were nineteenth-century figures like Walt Whitman, John Muir, and Thoreau, but we also know that those men felt the growing shadow of the Industrial Revolution, which was hugely accelerating the pace of specialization. An apt metaphor of that process is the Appalachian coal miner having to go deeper and deeper and see less and less daylight, say nothing of the vistas once visible from the ridges of the mountains where he lives. In parts of West Virginia, the ridges themselves were (and still are) lopped off. And on the ridges that have survived, visibility has been diminished.

Galen Sherer, writing for the website Hikers for Clean Air, recalls that a naturalist named William Bartram wrote, after his first hike to an Appalachian peak in 1776, "I beheld with rapture and astonishment a sublimely awful scene of power and magnificence, a world of mountains piled upon mountains." (For those not familiar with eighteenth-century English, "awful" in those days meant *awesome.*) Since then, according to scientists who have monitored air quality on the Blue Ridge in Virginia, average visibility from Appalachian ridges has dropped from ninety-three miles to twenty-three miles. And in New Hampshire's White Mountains, the trend is the same. Bruce Hill, Senior Staff Scientist for the Appalachian Mountain Club, wrote in 2011:

> From Mount Washington you should be able to see 130 miles. Last Memorial Day, one of the biggest hiking weekends, you could barely see to the valley [a distance of about 5 miles as the crow flies] below.

If William Bartram's awesome views have disappeared, it's in part because of a critical—now broken—link between our physical senses and our capacity for mental vision and planning. A tragic manifestation is the smog over China, where the smoke of coal-fired power plants, along with the exhaust from millions of motor vehicles, has reduced visibility in Beijing and other cities to near-zero at times. The *literal* inability to see even fifty feet across a car-clogged street has provoked—possibly too late—a confirmation, for many, of how catastrophically that country's leaders have failed to see (or perhaps care about) what happens a few decades from now. A likely eventual result, for the world's currently most populous country (India is expected to take over that dubious distinction, if it hasn't already), is unprecedented levels of asthma, emphysema, lung cancer, and a crippled economy—quite aside from the more general, international, collapse we should soon expect.

Are there still visionaries among us? A few, perhaps—but the storm surges of high-speed technological development have battered them, or driven them into self-imposed seclusion. Recall the miles of cluttered wreckage that was left along the New York and New Jersey shoreline after Super-storm Sandy in 2013—or across Moore, Oklahoma after the nearly two-mile-wide tornado that same year. Recall the monster Cedar Fire of 2003, which burned 2,820 houses and other buildings in Southern California. Recall the survivors who returned, in all these places and a thousand others in recent years, to the rubble where their homes had been. I doubt that they were reflecting on the big-picture significance of the mounting assaults and wakeup calls of climate change, but—rather—were very narrowly focused on looking for the family photos or grandmother's urn that might have survived the devastation. For a visionary today, contemplating our hurtling civilization and at the same time trying to deal with one's personal travails must feel a bit like those survivors' sad search for mementos of the hopes they once had.

For the Kids and Grandkids

The shocks to our ability to envision have proliferated, as technology has controlled more and more of what we see or know. But I'm thinking of several categories of debilitation that have been especially severe:

- **The imagination of a child**

  Everyone seems to agree that play is critical to a child's mental, emotional, and social development. A child deprived of good play would be at serious risk of becoming a bully, sociopath, addict, or—most significantly—child abuser himself. For a lot of harried parents, toys really are nannies. If a toy helps a kid practice his developing physical and mental powers, it helps the kid build his own independent capability. If it simply does *for* the kid what the kid still needs to learn to do with his or her own hands and brain, it impedes development. A tricycle helps the kid grow capacity by exercising his legs and giving him a sense of how a person can employ mechanized efficiency to increase his own powers without usurping them. It's a natural step toward riding a bike, and then safely driving a car. On the other hand, an electric-powered toy car in which the kid simply sits and steers actually steals a bit of the kid's childhood. I've occasionally seen a young kid—maybe six or seven—driving on a suburban street in a gasoline-motored toy car, and it saddens me.

- **Storytelling**

  Whether by a tribal elder recounting great legends, a summer-camp counselor telling ghost stories around a campfire, or Mom or Dad reading *Where the Wild Things Are* to a toddler, storytelling has been a key part of human culture and mental development since prehistoric times. For a long time, storytelling was oral tradition: the raconteur spoke, and his audience listened and imagined.

The shift to electronic media has been both wonderful and terrible. It's wonderful when the directors, actors, cinematographers, and all those other people you see in a movie's credits can work together to tell a story in ways that engage the viewer's imagination at deeper levels than the camera and sound alone can convey. While the viewer is not producing his own mental image of the characters as would happen with reading, the actors' facial movements, intonations, or hesitations may stimulate him to sense—or vicariously experience–emotional dynamics that would not have occurred with reading. Movies (or yes, TV shows or videos) can be among the greatest forms of art when they combine all their contributing crafts in ways that engage a wide range of emotional and neural responses. For those of the far-seeing who survive the coming collapse, good storytelling may be as essential a form of sustenance during the transition as a good food source.

On the other hand, simplistic media that don't offer such depth or nuance, but just relieve the viewer of imaginative work, tend to reinforce passivity. One example: simplistic crime drama replacing any engaging of moral wonder about why good people do bad things, or why good occupations are so often corrupted by bad intentions—or why some people are powerfully motivated by their visions of the future and others just aren't. But whether it's kids' cartoons, family dramas, or sit-coms with laugh tracks to help everyone know when to laugh, mediated storytelling—if too simplistic—can weaken imaginations and numb cognitive development in millions of young people.

- **Active entertainment**

Entertainment has undergone two epic transformations in the latest 0.002 percent of human evolution: it has sagged from active participation to mostly passive watching; and it has gone from being a tiny fraction of the average person's waking life to a share that rivals or exceeds the time spent working or sleeping. These changes are weakening for everyone, but especially for children. It's not just that kids like to play sports, dance, sing, pretend, or act—it's that those active entertainments build self-confidence, motivation, and anticipation that provide a foundation for healthy, resilient, and happy adult experience. When my grandson Josh was around eighteen months old, I was delighted to see him respond to music by doing exuberant, spontaneous dancing that looked a lot like some of the break dancing we'd seen on the TV show "So You Think You Can Dance"—a show that I'm fairly sure he'd never seen. There's nothing wrong with watching other people dance on TV, but if our nanny-techs make it easy for kids to spend four or five hours a day watching professional entertainers instead of entertaining themselves, they're being robbed.

- **Playing without toys**

Here again, I've learned something eye-opening from my grandson. When he was just over a year old, he began letting me know that what he loved to do, more than anything else, was to go out for a walk. "*Walk!*" he'd say (one of the first words he learned), and head for the door with a beseeching look. But a walk, for him, wasn't just a walk—it was an exploration that engaged every part of him. He'd *run* for a while, then jump, then stop suddenly, pick up an acorn, and say "*ball!*"—and roll it in his fingers. He'd point up into the trees and pronounce his version of "bird" (no "d"), and I'd look hard and finally see it—a tiny glimpse, far off, that I would not have noticed myself. He'd point at a cactus along the roadside and say "cacki, *no touch!*" He'd pick up a pebble and say "*rock!*"—and throw it with a surprisingly good overhand.

But most engagingly of all, when we came to a place where there was a dry dirt embankment, he'd stop, squat down, and pick up a handful of fine dry silt and let it pour slowly out of his hand onto the ground. Then another handful. Then he'd cup both hands and pour a double handful. He'd push the dirt along the ground with his hand held stiff like a tiny bulldozer blade. He'd dig a hole, using his hand like a tiny backhoe. He'd pick up a handful and throw it into the air. He'd find a twig, hold it between his thumb and forefinger the way you'd hold something with tweezers, and hand it to me as if he knew I'd been looking *all over* for one of these.

Watching the little boy play without toys, with just his eyes and hands, and the stones and twigs and dirt, was a kind of revelation. Later, I'd read about theories of education suggesting that an essential part of a child's development comes through the kid's own explorations with his hands and feet, well before his more cognitive forays into words and concepts. And then I thought of that anthropological theory of how human inventiveness began with the making of implements that extended the skills of the hands. It struck me that our modern rediscovery of the satisfactions of working with hands is a reconnection with the primordial skills that made us tool-users and inventors. It's a thing that one- and two-year-olds instinctively grasp, and I'm convinced it's a part of healthy development. For a kid who has a toy bulldozer, but no chance to play in the sand and dirt with his hands, there may be an important set of neurons that aren't getting activated. For optimal development of their inventiveness and creativity, kids need to spend at least some of their time toy-free.

Full disclosure: By the time Josh turned two, he had attracted toys from doting adults the way some empty lots collect trash. The toys (and pieces or parts of toys) piled up in his room, and sometimes it was hard to walk across the room without stepping on a few. Unlike when his Mom was his age, lots of Josh's toys can talk with high fidelity, and make startlingly realistic sounds. One evening I was making my way across Josh's room trying carefully not to wake him, when I stepped

> on something and suddenly I heard a man's voice saying loudly, "*Come out with your hands up! We've got you surrounded!*" I had stepped on a toy helicopter with the word POLICE on its side—and before I could catch my balance my shoe must have bumped another part of it and I heard the sounds of automatic weapons fire. This, for a *two-year-old*? I soon learned that kids today have toys that not only play *for* them in every conceivable way (Josh also has a garbage truck that can pick up a trash bin, lift it to the input bay, dump it, and return it to the ground), but can also induce a fair amount of phase-3 disconnection (recall Chapter 3). Now, when it's a good time for gift-giving, my own inclination is to skip the toys and give Josh a little of his grandpa's companionship—going for a hike, or letting him help me with a work project. By his third birthday, he'd had many hours of fun digging dirt with a trowel or mixing concrete with that same trowel, climbing a steep hill on hands and knees, and raking with a real (adult) rake. He has exhibited far more interest in my hand tools than in a lot of his toys. I did have to explain to him once that some of my tools are only for adults. He accepted that, but then smiled and said: "Grandpa, *I* a 'dult. *You* a kid!" I laughed, and liked it that he was able to defer instant gratification with good humor.

Einstein's Brain and Our Children's Brains: Can We Reconnect?

I don't know any of the people who design IQ and aptitude tests. But if the success of our leaders—in the United States and worldwide—is any measure of our intelligence as a species, we're in deep, deep trouble. The simplest way of describing that trouble is to observe that the vision of these leaders is riddled with disconnection—between body and mind, intellect and spirit, physical and virtual reality, sane and crazy, Christian and Muslim, New York and Texas. Ultimately, it adds up to a fateful disconnection between our human past and our future. And these are people—our glad-handing, gun-slinging, cowboys!—who have a big influence on what our kids will study in school.

If we can't learn from what's happened so far—in hominid evolution, Paleolithic persistence hunting, the agricultural revolution, the Ancient Greeks, the rise of Christianity, the fall of Rome, the spread of Islam, the Dark Ages, the Black Plague, the Renaissance, the Inquisition, the World Wars, Hitler, Hiroshima, the Cold War, Vietnam, The Exxon and BP oil spills, Iraq, ISIS, the killing of the western prairie, and the *World Scientists' Warning to Humanity*—we're not going to go any further. The journey will be over. And I'm determined that it *not* be over, but instead—for those who learn—that it be a new beginning.

Whatever intelligence may be, it's of little use if it won't help us survive. The great tragedy of our passing moment—the most recent

0.002 percent, let's say—is that our amazing brains actually have the capacity to make all those missing connections, or to repair the recently broken ones. And I have a hypothesis about that: miracles don't happen suddenly, as often depicted in religious literature, but over the course of thousands of years. The miracle of the human brain is that it has recorded and encoded critical lessons from the upheavals of history and prehistory. And with proper exercise, nutrition, and training during the life of an individual who *has* one, a well-oxygenated brain can connect all those epochal experiences in ways that let that person see a path forward. With clarified vision, we still have a chance—or at least many of us do—to continue the human journey.

A clue to this possibility appeared in the news in 2013, when the scientific journal *Brain* published a report of new findings about the brain of Albert Einstein, which was removed from its cranium after his death and divvied up among various scientists eager to do research. Since the dawn of the nuclear age, Einstein has had a reputation as the most intelligent person ever. We'll never know how Leonardo da Vinci, Marie Curie, Isaac Newton, Emilie Chatelet, Benjamin Franklin, Elizabeth I, or Christopher Marlowe (aka Shakespeare), would have compared with Einstein in IQ, but no matter—none of them had their brains posthumously sliced up and examined for posterity, as Einstein's was.

After Einstein died in 1955, and his brain was harvested by Dr. Thomas Harvey, not much of interest was found by Harvey or other researchers—until more than half a century later when *Brain* reported that researchers at Florida State University, revisiting the many photos taken in earlier investigations, found a simple but perhaps highly revealing fact: Einstein's left and right hemispheres were unusually well connected.

There's long been a large body of literature, both scientific and pop-psychological, about the separate functions of the two hemispheres—including the characterizations of particular kinds of talents as right-brain or left-brain. What follows here is unabashedly speculative, but.... The plasticity of our brains (which is well established now) allows us to form our own mental models of the world as we understand it. If our experience is increasingly compartmentalized, I wonder if that might make the coordinated functions of separate brain centers less fluid, and the walls between them more rigid. Maybe when the brain is at its most interconnected—literally open-minded—it is more able to create macrocosmic views of the cosmic world we move around in.

If we assume there may still be a chance for our children to grow up and save the world, even if we elders have failed, a hypothesis like that would be well worth exploring and might have significant implications for how our kids are educated. And in fact, an abundance of other evidence points in that direction. Interdisciplinary studies, virtually unheard of in the early years of the twentieth-century tech revolutions, have become prominent in the curricula of most top colleges and universities today. Yet, at the same time, the for-profit "tech" colleges that keep trolling for students on TV seem to have gone the opposite direction, more or less dumping the liberal arts—and courses that stimulate big-picture thinking—altogether. "Tech" rules all. Among the implications for twenty-first-century education:

- Academic disciplines that don't venture into interdisciplinary studies may have a deadening effect on kids' ability to see a viable path ahead. In the curricula of the future, interdisciplinary perspectives may need to be universal, and may prove as important as the core subject. In their book, *The Stork and the Plow*, Paul Ehrlich and Gretchen Daily wrote that the obliviousness of economists to the realities of planetary limits "traces primarily to a lack of basic education in the physical and natural sciences." They cite a 1987 survey of graduate students in economics, on the importance of other fields to their development as economists, who thought they didn't need knowledge of physics to be good economists. The lowest score was given to physics. Only 2 percent of the students considered it very important, while 64 percent rated physics "entirely unimportant." Those economists are today's leaders in the field, and their disregard of physical science still prevails—and helps to explain the still prevalent myth of perpetual economic growth that has contributed to the global breakdown we're now seeing the signs of. In the future, basic knowledge of the earth's material and energy resources—basic physical and biological science—should be a requirement for all economists.
- Specialization is obviously essential to safely managing technology in a heavily industrialized world, but because specialization alone can foster intellectual rigidity and myopia, it should be preceded in each young person's education by a broad education, *not* just because it's stimulating or "enriching" to take liberal arts courses or because it might help a student decide what to major in or what profession to pursue. *Broad education may be neurologically essential* to preparing for deep knowledge that can safely interface with the world outside its depths. Imagine, for example, the danger of having American health policy guided by a genetic scientist who has never studied the histories of colonialism, Nazism, eugenics, Rwanda, biodiversity, or some of the literary works that explore human conflict, yearning, or hubris.
- A general study of science—*what it is*—may be as essential to the education of every kid as is a general study of history and culture. In American schools, kids are taught what science has discovered, but

not much about the nature of discovery. They're taught the answers that science has given us, but not so much about the new questions it raises. And very little about the implications of scientific findings for their own futures. Meanwhile, thanks to a relentless media campaign by conservative investors and ideologues, public distrust of science has actually increased since the *World Scientists' Warning* and first reports of global warming. In 2013, a report by the *American Sociological Review* found that the percentage of self-identified conservatives who said they believed the earth was warming had *declined* from 50 percent to 30 percent in just two years—even as the scientific evidence of rapid warming piled up. Analysis of the polling suggested that many people believe science is a tool of liberal elites, and do not regard it as an unbiased means of understanding our world and ourselves. But it's not just ideological rigidity that has blinded so many people. The fragmentation of knowledge into separate fields, then into very narrow areas of interest within those fields, has made it harder for all of us to see how things are connected. Yet, at a time when understanding connections is critical to our long-term survival, we urgently need the whole population to grasp that science is not anti-religion or anti-Republican, but in fact is essential to figuring out how we are going to be able to eat, breathe, and survive in the coming years. In the age of rapid climate disruption and biodiversity loss (the sixth great extinction event), a general human respect for science is essential to societal survival.

- The kind of vision essential to survival is not just the ability to foresee future consequences of present actions (and to care), but also the insight that comes with active questioning of *why we are here*—the quest for meaning. That kind of questing is conspicuously lacking in American education, except perhaps in some parochial schools, which are not constrained by the sometimes too-simplistic separation of church and state. But "meaning" is as critical to secular life as it is to religion, and the bureaucracies of public education make a terrible mistake when they subordinate the methods of imaginative teachers—whether of literature, history, art, or the philosophy of science—to lockstep curricula. It's a shame, because children and adolescents have a natural proclivity for challenging accepted rules and norms, in ways that can greatly stimulate creative or inventive capability. But authoritarian systems of education suppress that proclivity, which then often ends up, in whack-a-mole fashion, popping up in kids' preoccupation with drugs, excessive risk-taking, and illicit adventures—or with third-phase disconnections from reality in the passive excitements of video games or vampire movies. If a kid wants adventure, or a sense of mission, the producers of "Pirate's Creed" or "Hunger Games" or "Grand Theft Auto" will be glad to provide *their* sociopathic versions of it—for a price, of course.

Some religious groups have worried about this lack of questing for meaning, and have tried to remedy that lack by teaching doctrine.

But for young people in a largely secular culture of entertainment and gratification, that kind of teaching has been largely shrugged off. With philosophical questioning getting no great stimulation from either religion or technology, the consumer economy rules. Kids, who should be sharp-eyed descendants of the persistence hunters, become what the kids themselves like to call "clueless."

One of the best illustrations I've seen of that sad trend is a Doonsbury cartoon strip in which the dean of a third-rate college tells the college's president of his concern about the fact that half of the students have "made no gains after two years of college," and that "they spend three times as many hours socializing as studying."

The president replies, "C'mon, Dean, that's why they come! And as long as we give them good grades and a degree, their parents are happy too! Who *cares* if they can't reason?"

The dean: "Uh. . . employers?"

In the last panel, an employer is asking a young man, "Any special reason you're late, son?" The kid responds, "Yes, sir. I got trapped in a paper bag."

So, quite possibly, the biggest need of education is to foster a new passion for serious quest, whether in school or at work—to encourage stepping away from the in-box, Xbox, smart phone, or Facebook page—and putting energy into finding out what it takes not to be "saved," but to actively save ourselves.

For the Far-Seeing

In the transition to a new beginning beyond collapse, education will have to displace the industries of war, defense, and "intelligence"—and the proliferation of aggression-themed entertainments those industries foster—as primary consumers of society's energy and attention. Children in the dominant nations today are widely subordinated to the demands of parents' careers and the funding priorities of politicians who are financed by lobbyists and political action committees that have no financial interest in children. Children have no voice in the US Congress or the Mexican drug cartels or the Taliban. For those of us who hope to be part of the protectorate that shepherds our species forward in the next few generations, one of the greatest challenges will be to shepherd our children above all other considerations—not to indulge or "spoil" them, but to give them the discipline, strength, and vision they will need to continue the journey.

How to go about that? If you are among the far-seeing, you've had the good fortune to encounter—and draw from—a wide spectrum of human experiences and skills. Now you have an opportunity to use what you see in a once-in-a-thousand-lifetimes way.

As you connect with others, in the emerging protectorate of the human future, re-think your relationship with children. Recognize that kids are not inferior to adults, just less experienced and skilled at what the present generation has considered important. But kids can also bring fresh vision to situations we older people long ago learned to treat with preconceived expectations and mental filters. (The inventor Buckminster Fuller said, "Everyone is born a genius, but the process of living de-geniuses them.)

As you form your community, look for ways to form mentoring relationships with kids, in which you teach but also consciously learn. Allow ample time for reading, storytelling, art, and play. Also spend time practicing survival skills: getting to know the native wildlife and ecology in the place where you live, and identifying the possible uses and reuses—at a sustainable scale—of local resources.

Consider the possibilities of home schooling. Ideally, the United States would give vastly greater support to public schools than it does, but the constraints of underfunding, political and policy battles over testing and curricula, and the bright-flight to private schools have left public schools foundering—one of the early signs of societal disintegration. As the bureaucracies of education ossify, far-seeing families will need to take over their kids' futures.

10

Hands, Feet, Bare Skin, and Privacy

The Last Bastions of Natural Strength

In his book *The Shallows,* Nicholas Carr writes that our intelligence is under assault. I couldn't agree more. But the implication that the thing we need to defend is our *brains* may be short-sighted. It's like the view that when American democracy is threatened, it's our *territory* (our "homeland") that must be defended. To defend democracy, what we really need is to rebuild the respect for literacy and education that fostered the democracy we inherited—otherwise the territory is no more special to us than that of Russia or China. In the decline of our big-picture perceptual capacity, not just as Americans but as a species, it's not our brains that need defending, but the amazingly designed and coordinated parts of our physical and sensory experience that made those brains so robust to begin with. Without that experience and those parts, the brain becomes aimless. Without our evolving hands, feet, bare skin, bipedal anatomy, and wide-ranging exploration of the earth, the human brain would have remained like that of a chimp or gorilla. This observation might sound like a foolishly ageing-hippie, anti-tech longing for pre-modern existence. It's not. Rather, it's a reality check about how the brain first evolved and then eventually—in epic irony—enabled the misuses of technology that now weaken it.

For the hands to play concert piano or the feet to play NBA-level basketball, or for bare skin to allow a marathon-running man or woman to cool the body enough to keep it running for five hours at a speed that a majority of high school kids today couldn't keep up for five *minutes*—to do any of those things, the brain had to grow. For many millennia, that growth was not yet about things like using the fingers to play the music of Mozart, but about the very delicate work of shaping flint cutting tools or spear points. It wasn't yet about running athletic patterns on a basketball court, but about running over primitive trails—dodging snakes, rocks, roots, holes, and thorns, chasing down animals for food. It wasn't yet about planning next year's budget, but about tracking a route through a complex wilderness that would make or break your ability to bring forth the next generation.

During the agricultural revolution of five to ten thousand years ago, those hunters' descendants lost much of this daily contact with the wild earth. In the last few centuries, with the industrial and then atomic, biotech, and digital revolutions, the disconnection has become critical. The proof is in the unhesitating inclination of economists to teach their students "laws" of economics that make no reference to the physical reality of the earth.

The Ungrounding of America

Now, no less than a hundred millennia ago, the life we live springs *from the ground* in the form of plants we eat (grains, fruits, and vegetables) for metabolic fuel, ground water required by those plants and by us, or oxygen emitted by those ground-based plants or the oceans. Life also thrives in the forms of animals that eat those plants and drink that water, or that consume other animals that do. Take away that ground-based environment and what it has produced, and you couldn't live another ten minutes. Yet, in the United States today we have thousands of public officials who call themselves "conservatives" yet regard the environment *not* as something to *conserve* for the survival of all, but as a "special interest" of just one political group. Their conservative ideology doesn't seem to include conserving the earth, water, and air that keep them and their loved ones alive. Willfully oblivious to what they depend on for every breath they take, they are the epitome of *un*grounded.

In Britain, recent research suggests that there may be a genetic basis for this political divide. The National Longitudinal Study of Adolescent Health found that subjects who identified as "very conservative" tested significantly lower in IQ than those who identified as "very liberal" (hold on, now, before you burn this book), and the difference was explained by evolutionary psychologist Satosh Kanazawa of the London School of Economics as a difference between biological traits that are "evolutionarily familiar" (such as distrust of other clans or tribes) and those that are more recent and more adapted to the rapidly changing conditions of a world in which successful coexistence and cooperation with genetically unrelated strangers is critical to survival. I don't think much of IQ testing. But Kanazawa's conclusion has to make you wonder, what *does* make so many of our fellow Americans seem so eager to win Darwin Awards (recall Chapter 1).

We humans evolved with our feet on the ground literally, and on an epic scale; we migrated by foot over the whole earth and inspired the legends of great journeys, in native cultures the world over—and then the real-life odysseys of Marco Polo, Magellan, Lewis and Clark and Sacagawea. But over the past two centuries, in reactions to the Industrial and high-tech revolutions that have increasingly separated us from our earthly connections, ripples of protest or apprehension have emerged: the nineteenth-century Romantics, Thoreau's civil disobedience; John Muir's rescue of wilderness; the "good life" of Helen and Scott Nearing; the 1970s "back to the land" movement; deep ecology; and Rachel Carson's *Silent Spring*. Whether via subconscious genetic memory or the emerging sciences of ecology, the environmental movement that emerged from those spurs reintroduced hundreds of thousands of us—including me—to our planet.

In 1969, my brother Bob was doing research on the possibility of replacing the internal-combustion (gasoline) engines with electric motors in American cars. He met with Ralph Nader, author of the blockbuster *Unsafe at Any Speed*, to discuss coauthoring a book that would move beyond the safety problems of cars to investigating their environmental and cultural impacts. Within months, it became apparent that both Bob and Ralph were up to their ears in other projects. Since I was already working with Ted Taylor and Bob at their firm IR&T, the job was handed off to me. I knew nothing about cars—couldn't even change my oil. I was just an editor, but apparently doing an OK job with Taylor's *IR&T Nuclear Journal*. So, somewhat naively, I took it on.

Nader put me in touch with the US Senator Gaylord Nelson of Wisconsin, who was conducting hearings on the auto industry, and who was glad to share the transcripts. I titled the book *What's Good for GM*, in sardonic reference to a comment by Charles Wilson, who'd been chairman of General Motors when it was the largest, most powerful company in the world. "What's good for GM is good for America," Wilson had famously declared. At the time I took the job, GM was still the goliath of American industry. Senator Nelson wrote a foreword to the book, and the following spring he joined the environmental activist Denis Hayes to organize the first Earth Day. The book didn't make a ripple, and its biggest moment came when the publisher paired it with Daniel Schorr's book *Don't Get Sick in America*, in a display ad for the *Wall Street Journal.* Maybe it wasn't a propitious time to be challenging the hegemonies of either the auto industry or the medical establishment, though. There were no reviews and no more ads, and shortly afterward the publisher disappeared.

Still, to be in my late 20s at the time of the first Earth Day was exciting—a tug-of-war between youthful hope and creeping anxiety about the industrial exploitation of my world. I loved running on forested trails, but worried about what the fate of the forest might be. I bought copies of the first *Whole Earth Catalogue*, and the first *Mother Earth News.* And then I bought a five-acre lot adjoining the west slope of the Shenandoah National Park, a couple of miles below the Blue Ridge in Virginia—a little beyond an atomic-bomb's immediate reach from Washington. The lot had a spring, fed by wild uphill forest, and my plan was to build a cabin that could be a retreat if things got bad. It would be my safe-house castle, nearly half a century before the TV show "Doomsday Castle" would appear. More on that a little later.

My Cabin of Stone

The way I built that "castle" might be considered a bit Luddite, and I know I wouldn't do it quite the same way if I were doing it today. And in fact I probably wouldn't do it today at all, because I now understand that an isolated, ultra-self-reliant existence is untenable and quite contrary to human nature, and the only path to survival is through community and cooperation. But looking back, I don't consider the experience to have been a mistake. Rather, it was a path to epiphanic discovery about the connections between my hands, legs, lungs, and brain, and it proved to be a course I could never have taken in any classroom.

I'm especially glad I had this experience before the era of email, texting, and social media. It was profoundly good not to have distractions. In addition to the spring, the property had a small creek running through it, gathering small rivulets and streams from the forest above. Centuries of spring runoff and summer rains had tumbled thousands of rocks, and the creek was full of them. I'd been an avid rock collector in my teens, so finding this creek was for me like discovering King Solomon's lost mine, with the legendary gold all still there. I was entranced by the variety of rocks in the stream-bed: granite, limestone, puddingstone, feldspar, basalt—and as I looked closer, sometimes a pocket of quartz or amphibole crystal, or glittering slab of mica schist. The creek was about 200 feet from the site where I wanted to build my cabin at the edge of a meadow backed up to big trees, and I decided to build the cabin of these stones.

Of course, I wasn't a real Luddite, because I commuted from Washington to the Shenandoah on weekends, driving my 1966 VW Beetle convertible, and after I'd hand-dug (with pick-axe and spade) a perimeter trench for the footing, I hired a concrete company to come up the gravel access road and deliver a truckload—nine cubic yards. On the day the cement mixer arrived, I witnessed a dead metaphor coming back to life as the man pulled out the chute and filled my trench with the whole nine yards. Then I had the company deliver a dump-truckload of sand, so I could hand-mix my own cement in wheelbarrow-sized batches to mortar the stones. On every weekend trip to the site, I stopped at the nearby town of Elkton, where there was a building supplies store, to buy ninety-pound bags of Portland cement, rebar, and any tools (spade, hoe, trowel, buckets, spray bottle) I'd need.

After that, though, it was nearly all hands, arms, legs, and back—and eyes. I'd hand-carry rocks (most of them between 50 and 150 pounds, with an occasional, beautiful 200-pounder for the ages) from the creek, or haul them in a wheelbarrow, then go back with a bucket to dip water for the cement. Mixing three shovelfuls of sand to one of Portland cement, and enough water to achieve the consistency of mashed potatoes, I'd go to work. Each rock I'd select for both fit and appearance. I wanted to be sure it would fit securely in place without wobbling or falling even without being cemented; then I'd remove it, place a bed of cement, and replace the rock in a position where I expected it to remain for the next 200 years. I didn't want a single rock to depend on the cement to hold it in place. The cement would be entirely redundant strength.

Once the rock was cemented, I used a trowel to rough-shape the seam, then my fingers to smooth and seal the joint. At first I used cotton work gloves to protect my fingers from the caustic chemical in the cement, but the tips of the glove fingers quickly wore through and after a while I said to hell with it and used my bare fingers. I'd go to the creek and wash my hands, and the tips of my fingers would be wrinkled like overcooked peas. If it was a warm, sunny day, I'd come back to the rock an hour later to spray a mist of water over the seam—and over all the other work I'd done that day or the day before, as well as on the spot I'd selected for the next stone.

The walls of my cabin were fourteen to eighteen inches thick, and always in the back of my mind was the thought that the cabin needed to be strong enough to last centuries. Why? I wasn't yet aware of the "Seventh Generation" creed, but I think the sense of slow collecting and shaping of rocks I felt with each trip to the creek, and the imperceptibly slow growth of the trees behind my walls, eased me into a slow rhythm that felt in sync with my surroundings. I had no earbuds or phone to distract me. Even the Walkman was still in the future. The most satisfying part of the work was finishing each rock, then standing back and gazing at it, savoring the sight the way I'd occasionally savored a bite of some especially good food. It's hard to explain why, but that stone-by-stone accretion of satisfying contemplation was as character-building for me as it was wall-building. Never mind that my fingers would feel as if they'd been sandpapered. I was pulled irresistibly by the slowly changing look of what I'd done. Building the walls took about two weekends a month for seven years.

Lifting and carrying those rocks made me strong for my fairly slender build. I'd drive to the Shenandoah on a Friday evening, camp out, and work from first light the next morning to about 4 PM, then go for a long run. Sometimes I'd run up the road to Elkton, then up the long climb to the Blue Ridge, where the road crossed the Appalachian Trail. This was years before the Utah and Harvard biologists found their astonishing evidence that humans had evolved as long-distance runners, but intuitively I felt very close to my nature. In the sixth year of my cabin-building, I decided to enter the iconic JFK 50-Mile endurance run, in the mountains of western Maryland—America's largest ultramarathon—and in my training runs I envisioned winning it. And then on the day of the race, the dream came true. It was the most thrilling six hours of my life.

The rock-hauling, lifting, and positioning—and the shaping of the cement as if it were marble or bronze—had put me in touch with my hands, and the running had put me in touch with my feet. The long hours I spent running, in training for long races, had also put me in touch with something else: the magical feeling of a strong heart and bare skin in open air. Weather permitting, I'd always run without a shirt. Later, I'd learn from exercise physiologists what was happening: the human body, if working hard, builds up heat just as an engine does. If it overheats, like an overheated engine it fails. As long-distance persistence hunters, humans eventually became more naked because bare skin cools the body (via convection, radiation, and the evaporation of sweat) much better than a skin covered with fur or hair can. Bare skin felt good—a legacy of its evolutionary advantage. If being naked feels good for many of us (and might for all, if cultural impediments didn't intervene), it's not just because of our association of nakedness with sex.

When I was building the cabin, I didn't feel either guilty or hypocritical about driving long distances and consuming gasoline (and hiring big diesel trucks to deliver concrete and sand) in order to have the experience of building a stone cabin without power tools. It wasn't the rejection of technological aids that gave me a kick, but the feeling of having my hands, feet, and half-naked body get close to the experience of being the wild animal my not-so-dumb ancestors were. It wasn't a negative impulse but a positive one—a kind of love. E.O. Wilson would call it biophilia.

One day I was working at the cabin, mixing cement with a hoe, when a very large black bear walked out of the woods into the meadow about a hundred feet from me. It stood up on its hind legs, presumably to get a better look at me over the tall grass, and I became a good Quaker and stood very still. For a moment there was a delicate mutual inspection, and then the bear turned and walked back into the woods. If I'd been working with a noisy gas-powered cement mixer, I don't think the bear would have approached. In the four decades since then, I've seen many signs that the satisfaction of working with my hands gives me a stronger connection to nature—and to *my* nature—than working with labor-saving devices does. And I know I'm far from alone. Hand-built houses, craft fairs, community gardens, and art galleries have a fascination for at least a small percentage of people who know that what can be experienced in these places cannot be matched by high-volume mechanized production. The word "artisanal" evokes not just higher quality in a growing range of small industries, but a lifestyle to which growing numbers of us aspire. Beyond the consciousness of that yearning may be an instinct to recapture the kinds of work and play that nurtured the growth of our brains and capacities, and to fend off the tech incursions that threaten to reverse that growth—or more likely have already been reversing it for decades.

Fortunately, at least a few of our more independent thinkers have not forgotten that critical connection we're in danger of losing altogether:

> "Heaven is under our feet, as well as over our heads."
>
> –Henry David Thoreau

> "Stretching his hand up to reach for the stars, too often, man forgets the flowers under his feet."
>
> –Jeremy Bentham

> "Keep your eyes on the stars, your feet on the ground."
>
> –Theodore Roosevelt

> "Forget not that the earth delights to feel your bare feet and the winds long to play with your hair."
>
> –Khalil Gibran

And then there's this, from the "Peanuts" strip by Charles M. Schulz:

> "Jogging is very beneficial. It's good for your legs and feet. It's also good for the ground. It makes it feel needed."
>
> –Charlie Brown

For the Far-Seeing

To transition from the quest for a viable path to actually setting out on the journey, you'll need to move with high alertness, acute peripheral vision, great patience, and endurance—and maybe eyes in the back of your head. In other words, like one of our prehistoric persistence-hunting-and-gathering ancestors (only the best of them *became* our ancestors), you'll need extraordinary skills of a kind most people now don't have. If you sense that a key need now is to train your brain for what's ahead, don't be distracted by the many "brain power" exercises appearing in popular media. Those are games. What builds the capacities needed for our journey is hard work, especially work that calls on a range of faculties—physical, emotional, perceptual, cerebral.

Suggestion: pick a project or two—building an off-the-grid house, planting an organic garden, training for a marathon (not one you'll just jog to finish, but one you'll really run!), making a film, reading ten great books, mentoring a kid who's very troubled but bright. As you work, sense the world with your hands and feet, as well as with your eyes. And then consider the possibility that to be truly visionary—to see well into the time to come—means studying the vast neglected realm within yourself, as well as what lies beyond the horizon.

11

Overriding Evolution

A Course of High Adventure and Uncertain Salvation

Technology not only builds on biological evolution but also overrides it—the "higher" the tech, the faster the override. Biological evolution has sometimes come in spurts, but compared to the pace of civilization, even the spurts are slow. In the sprint culture that has become dominant over the past half-century, the overrides of evolution have been recklessly fast—and flawed. Technology by definition takes us where our bodies and brains alone can't. But that also means our bodies and brains can't easily put the genies back in their jars.

When I went to work for my physicist brother Bob and his partner, the atomic-bomb builder Ted Taylor, as mentioned earlier, we wrote a mission statement that somewhat simplistically began, "Technology is neither good nor bad. . . ." By extension, what we wrote about the moral neutrality of technology could also have applied to the use of it to override evolution. But when an override occurs rapidly and with little regard for unanticipated consequences, it's opening the gate to trouble on a Biblical scale.

Of course, there have been specific instances of override that some of us might feel are *prima facie* wrong—notably human cloning, or building a superior race in the way that Hitler planned and that some of the futurists I've met still seem to envision. But again, it's the misuses of cloning or gene splicing, not the tech itself, that will be blamed if things go badly. And in any case, any argument about the morality of overriding evolution is now moot. We are too far down that very rocky trail to even dream of turning back—and in fact, we're now in a place where if we want to survive, we may have to take the overrides even further.

The Changes So Far: Inadvertent and Probably Irreversible

The history of civilization can be described as the history of technology—the magnification of our anatomical powers and eventually of the biochemical and electrochemical powers of our brains and nerves, thousands-fold. So, this history can also be described as the history of invention, and of our emergence as an inventive species. But technological revolution has also brought social revolution—first the great transformation from hunter-gatherers to farmers, and from experts in survival to experts in specialized occupations, but then also in our awareness as emotional beings, as *sapiens*. We invent but we also feel, with those often wildly conflicting capacities emanating from different parts of our brains and bodies. In the eyeblink that has been the last 0.02 percent of our biological evolution, we have experienced a heightening sense of ourselves as individuals with personal, as well as cultural, needs and aspirations. We've been caught up in storms of sexual, religious, and ideological passions that have changed our emotional landscape far faster than evolution can—and we've responded by devising ever faster and more powerful ways of satisfying those passions: new methods of birth control (a condom for women, a pill for men); new means of reducing weight; new means of spying, stealing, hacking, deceiving, or killing; new ways of getting rich or getting high. In the economics of invention, there's no time for testing the long-run effects of these inventions either on individual health or on the stability of society. They are the business of everyday commerce, and because they become obsolescent almost as quickly as they appear, their cumulative effect is to override evolution in ways that could never be measured even if we *weren't* racing time.

But a few of our self-administered upheavals are now visible enough to make it clear that stopping our drive to override nature is probably no longer an option, if it ever was—so the only viable way forward is to override with far greater deliberation and attention to future consequences. Some of the more conspicuous overrides of the past 0.02 percent of our time on this planet so far illustrate the point:

- **Sedentary heart**

  Hunter-gatherers were on the move as long as weather permitted and sometimes when it didn't. Even women with young children may have spent a lot of time walking in search of edible plants or small animals. Though humans take fourteen or fifteen years to develop most adult capabilities, we are much quicker to develop bipedal mobility. In the 1980s, my magazine *Running Times* published a profile of a three-year-old girl in Baltimore who went to a park with her father each week and ran ten miles. More recently, I went for a walk one day with my grandson Josh, who was not yet two, and when we got out to the street he began to run. Josh had never seen *me* run, and I had never encouraged *him* to run; it was entirely spontaneous. To my great surprise, he ran about half a mile nonstop, laughing and clapping his hands and raising his arms like a football player who's just scored a touchdown. He "talked" enthusiastically the whole time, mostly in a toddler's language I could only partly understand. (His verbal skill at this point lagged far behind his physical mobility.) The whole time, I noticed, he was taking about three steps per second—just about the same tempo you see in a good adult runner. (If the speed he was going was somewhat slower than an adult's, it was because his legs were a lot shorter, not because his tempo was any slower.)

The instinct to walk and run is very much with us, and our hearts and cardiovascular systems have evolved to support that activity. But in the most recent 0.4 percent of our evolution, phase-1 separation (fixed settlements and more sedentary occupations eliminating hunting and gathering) took away a lot of the work the heart and lungs evolved to do. The recent trend toward increasing phase-2 technologies (cars, telephones, telescopes, TV news, and Internet) and then phase-3 techs (virtual realities achieving full disconnection from the physical world) makes it clear that there will be no full reversal of this override any time soon. The sports of long-distance running, soccer, bicycling, hiking, and mountain climbing will compensate for some, but others are caught in a widening chasm between what they evolved to do and what they're not doing because nanny-techs are doing it for them.

- **Accumulating fat**

When hunting and gathering was curtailed or halted for prehistoric humans—sometimes for days or months, by ice-age winters or by losses of good hunting grounds to migration or disease—survival favored traits that allowed for storage of energy-rich fat in the body. Early humans may have had to hole up in caves or huts, and subsist on their own body fat until better hunting days returned, almost the way hibernating bears do.

Those traits are still in us (E.O. Wilson has said our genome is still essentially what it was 100,000 years ago). But modern technology has eliminated our exposure to long spells of seasonal food deprivation. Instead, we have millions of people who are as sedentary as winter cave dwellers may have been, except that they're sitting on the couch in a warm house or apartment eating fat-saturated potato chips or bacon cheeseburgers. No longer needed to survive the cold, the fat piles up on the belly and hips—and in the form of arterial plaque.

Countless override-the-override technologies have been devised to deal with that—from diet pills to lap-band surgery to treadmills. Some of them (such as botched lap-band operations, or heavy dependence on drugs) have had unanticipated, often fatal, consequences. Our weight-reduction methods have a long, long way to go to be as thoroughly debugged as our original physiology has been. The obesity epidemic is a result of being seduced by hundreds of new technologies, from home heating to food processing, into disregarding and overriding key aspects of our inherited nature. Deploying technological overrides to attack the cultural override has gone badly.

- **The herd no longer getting routinely culled**

This is touchy, but bear with me a minute. When researchers can identify genetic defects in an individual, medical technicians may be able to improve that person's otherwise predestined chances. It's an exhilarating capability, very close to the threshold of God's work of Creation, some might say. But there's a rule of inverse moral cost here: the more the risk of an illness is reduced by this means, the more the risk of medical hubris is increased. Civilized humans have a long history of building walls to thwart enemies: the Great Wall of China, Hadrian's Wall, the Berlin Wall, the steel-reinforced walls of the Pentagon, and the seawalls protecting the Fukushima nuclear plant, among others. Now we have the biggest wall of all—the phalanx of medical and pharmaceutical technologies that stand as a barrier against the evolutionary processes that weed out genetic defects and weaknesses.

The good news is that millions of children are protected from polio, tuberculosis, or the social abuses (including infanticide) once inflicted on children who are physically or mentally handicapped. But the long-range effect is also to stop the strengthening influence of evolution in its tracks. And because public memories are short, we too easily forget that as the pressures on them mount, rigid walls almost inevitably are breached. An article in the *Journal of Coastal Research* reports that sand dunes protect the shores better than sea-walls, because they are *not* rigid—they can shift, erode, and re-grow. The med-tech wall can't indefinitely withstand the pressures of population expansion and cowboy-economy depletion of resources. In the meantime, though, its effects on evolution are inevitable. We want the sweeping diseases of the past to be warded off, but need to be aware that if the med-tech wall breaks, our evolved protections will no longer be at full strength.

- **Humans becoming domesticated**

Homo sapiens is no longer a wild species: ten millennia of cultural breeding has changed our behavior and nature in countless ways. We now act in some ways more like cows, sheep, or house cats than like those animals' wild ancestors. We use the words "cowed," "sheepish," and "scaredy-cat" not to describe domesticated farm animals or pets, but *people.* (I haven't actually heard the term "scaredy-cat" used in a long time, but when I was a kid we used it a lot, mostly to tease or insult other kids.) In recent centuries, *most* of us have become cowed, sheepish, or scaredy cats. Even when massive public protests or uprisings, like the American Revolution, Vietnam War protests, civil rights marches, or the Arab spring of 2012 occur and the news coverage seems to depict an angry public aroused to action, it has usually been, in fact, only a small minority going to the streets or public meetings. And even if the protesters gain power, the gain may be lost if the majority who didn't protest remain passive.

Domestication is a broad, multifaceted change that social scientists have only begun to fully sort out. One theory is that biological domestication has made us less aggressive, less inclined as individuals to fight or kill other individuals over personal conflicts or just because they are different. That theory gets some support from studies of the domestication of other mammals, which found that aggression declined over many generations of selective breeding. The hypothesis here is that with the advent of civilization and the division of labor, the need for cooperation within a society became paramount. The blacksmith might be a very strong man, but he'd go hungry if he couldn't do business peacefully with the farmer or butcher—trading tools for grain or meat. Over-aggressive individuals were more likely to get killed off in fights or by punitive public sanction, and therefore less likely to pass along their aggressive traits. So, the upside of being cowed, sheepish, or scared is that you stay under the radar, stay out of bar fights or battles with the cops, and are more likely to survive and have kids who survive. But the downside is that the more passive and compliant most people are, the more likely it is that they will be manipulated or exploited by the few who remain most aggressive without being stopped—and the more likely that they will reinforce the passivity of the next generation.

A quick review of the overrides listed above strongly confirms the observation that the effects of cutting ourselves off from the biological shaping brought by the first 99.6 percent of our evolution can have both wonderful and terrible impacts. The perpetuations of fat accumulation and sedentary hearts are great concerns. The ability to eliminate genetic defects is in principle a wondrous achievement. If the challenge to society were merely a matter of picking which overrides to undertake and which not, it would still be daunting. With genetic engineering, for example, where do you draw the line between eliminating birth defects and designing a eugenically superior race? How do you respond to the US government's quiet plan (as acknowledged by DARPA) to create immortal organisms, a biological and ethical Pandora's Box if there ever was one?

But the challenge is not just a matter of picking good versus bad overrides, because every override entails tradeoffs with other, often unforeseen, impacts—a fact that the Creator of slow biological evolution evidently understood, but the champions of sprint culture have not. For example, consider the cultural transformation, with the rise of civilization, from the necessary aggressiveness of early humans who had to fight predators while being predators themselves, to the presumably more cooperative and less fight-prone breed we are (as individuals, not as gangs or nations or cutthroat corporations) now. If the result is that a cowed individual is less likely to get killed by a jealous boyfriend or punitive jury than a person who is prone to impulsive violence, so far so good. But it also means that if many more domesticated humans are now able to produce offspring before they die, that tamed condition becomes another form of discounting of their grandchildren's future: one benefit now (a less aggressive nature) contributes to three disbenefits in the future: more population, more propensity to being exploited, and more rapid exhaustion of the planet's natural capital to pay current expenses. What once helped the species to survive now weakens it.

That tradeoff between probabilities of individual survival and societal survival is part of a more general problem of disconnection—between the narrow view and the big picture. We are far beyond the point where we can reconsider the wisdom of overriding evolution, and now have no choice but to override it even further. Rather than back off for fear of further catastrophic mistakes, we'll have to continue moving forward, but with far greater attention to that long-neglected big-picture perspective. The future of technologically modified evolution needs to be guided not by the biological objective of making the human population larger, as it once was, but rather of gradually making it humanely smaller, with the prospect of a far more meaningful existence for each surviving person.

Here's where robust technology can play a heroic role, *not as nanny but as wise elder.* By the metaphor of the wise elder, I mean being programmed not just for the gratifications of the moment, but for a long view of both past and future as essential context. The technologies of civilization gave us vastly expanded capacities for memory, first in writing and then in printing and libraries, and now in that amazing Kindle or iPad, not to mention the veritable pantheon of Google and YouTube and Wikipedia. And now, in the linking of our computers, cars, phones, home security, and credit cards via the IOT, which is in turn linked to the NSA. Maybe that legacy of ever-expanding might was inevitable, as for all of recorded history and perhaps long before it, humans have looked upward—to the stars, the sun, the heavens—seeking not just quick enrichment, but legacy. But take another look at the five quotes at the end of Chapter 10. For those of us who dream and aspire, the trick—as demonstrated by great ski jumpers, dancers, or basketball players—is to be able to both reach for the stars *and* stay on our feet. With the technology we have now, there's momentous new work to be done—work that, the pitfalls of the sprint culture notwithstanding, 'twere best done quickly.

In for a Penny, in for a Pound: The Overrides We Need Now

The only way forward is forward. Life, by biological definition, can't come to a standstill or it slips into death. A civilization that comes to a standstill collapses, as surely as will one that out-sprints its oxygen.

As mentioned in Chapter 6, hard scientists describe this as the principle of entropy. All things disintegrate—*all* things, from apples left too long on the tree to entire societies—unless hard work goes continuously into shaping their futures against the tide of disintegration. A ripe apple can be picked and made into apple pie before it falls from the tree and rots. A society can try to remedy its weaknesses and build new strengths. A species must continue its adaptation, as thousands of other species around it do, or go extinct. In the latest 0.002 percent of our evolution, our technologies have wrought enormous impacts on the planet's ecological balances: more carbon dioxide in the atmosphere, warmer temperatures in both air and oceans, disappearing bees (which we depend on to pollinate many of our food crops), hundreds of radioactive hotspots that will be lethal to nearby life for centuries, and on and on. In evolutionary time, those changes have been too fast to allow many species to adapt, and that speed of change—along with our own expanding population and destruction of other species' habitat—is why we are now in earth's sixth great extinction event.

In the last such event, the Cretaceous–Paleogene extinction believed to have been precipitated by the planet's collision with a large asteroid, most of the dinosaurs and about three-fourths of the world's other interdependent species were wiped out. This time, it could be us. It goes against our waking experience to think so; we live our entire lives in that eyeblink of evolutionary or geological time, but our moment has also been a moment of explosive destabilization. Our only chance to avoid at best some form of civilizational collapse (recall Chapter 8) is now to use those big brains we've been so stupendously misusing and move from the largely accidental overriding of evolution we've done so far to a very deliberate and integrated one over the next few years. It's the ultimate gamble, but an existentially necessary one.

To Richard Dawkins, the renowned and often provocative Professor of Public Understanding of Science at Oxford University, it is appropriate now not to regard our deteriorating prospects as causes for desperation, or giving up, but as opportunities to get some critical things right that in recent times have gone wrong. "I am very comfortable with the idea that we can override biology with free will," Dawkins told an interviewer after the publication of his book *The Selfish Gene:*

> We must understand what it means . . . to be programmed by genes, so that we are better equipped to escape, so that we are better equipped to use our big brains, use our conscious intelligence, to depart from the dictates of the selfish genes and to build for ourselves a new kind of life, which as far as I am concerned the more un-Darwinian it is the better, because the Darwinian world in which our ancestors were selected is a very unpleasant world. Nature really is red in tooth and claw.

To that, though, we might do well to add a caveat: that while nature is indeed red in tooth and claw, we humans—because we never *had* sharp tooth or claw—are now red in the wounds of war and genocide and the reckless tech-empowered decimation of life in the future as well as in our own time. To override our demons, both genetic and cultural, we'll need to move with unprecedented deliberation—training ourselves even as an Olympic athlete trains, first envisioning the fully integrated performance we aspire to, then breaking that complex skill into components for studied practice before practiced reintegration. We'll need *both* specialized knowledge and big-picture perspective. And it's critical, now, that we avoid either the rush of the sprint economy or the hubris of those futurists who would turn us into cyborgs and move us to Mars, without ever having understood the planet under our feet.

To pursue that analogy of the athlete training for the performance of his life, here are some of the key components:

- **Rethinking how children learn**

  "Education" as we now know it puts heavy emphasis on providing answers, based on the presumption that teachers and institutions have already determined what those answers are and need only to drill them into the kids. But the message of the earth to its human occupants now is that too often we have *not* had the answers. In some critical areas, we haven't even been close. Real learning puts the emphasis on asking new questions.

  How does this override evolution? Prehistoric children, beyond what they learned from direct experience of their environment, must have learned everything they knew from their elders, since no major technologies of knowledge accumulation had yet been invented. That arrangement worked best if the children had confidence that the elders knew what they were talking about. An adolescent boy going on his first hunt for mammoth could do so with more of that critical confidence and decisiveness essential to success if he respected the older men who taught him and didn't ridicule or ignore them as kids so often do now. It also made for a more stable society if the elders were not too easily challenged.

That pattern carried over into historical times as well. Both the personal ambitions of leaders or teachers and the institutional strength of the community (or state, or church) seemed, at least to those in control, to be best served by authoritarian views of what's right or true. But the perpetuation of fixed truths, while serving well for people in small clans who had to survive the ravages of high mortality, does not serve well in a huge population racing toward catastrophe if it can't shift to a new course. Authority can be a heavy foot on the gas pedal, in a car-chase toward the cliff, when what's needed now may be more akin to Robert Frost's coming to a fork in the road and pausing to consider whether a road "less traveled by" may be a better way to go. Traditional education gives us a map, or a GPS voice you can't have a real discussion with, telling us which road is correct. *Learning* means *questioning* which way is best, and why.

It's hard to imagine educational establishments learning to respect students' questions as much as textbook answers. But there's something even harder that may be essential: recognizing that the old can learn from the young, as well. Childhood and youth sometimes have a clarity of vision that grows dim as the years fly by. It's the young who best embody that magical balance between seeing the stars and keeping grounded. As we grow older in our world of omnipresent violence, loss of vision, and heavy tech-dependence, we too often give up on our hopes. That giving-up was already evident in the throes of the Industrial Revolution, when many people apparently felt more yoked than liberated by their new powers. One day in 1854, Henry David Thoreau wrote in his journal:

> The youth gets together his materials to build a bridge to the moon... and, at length, the middle-aged man concludes to build a wood shed with them.

A century later, president Kennedy made Thoreau's imagined youth's dream a reality. Now we've reached a point where we really *need* our kids to be a century ahead of us—not by inventing new techs for digital empires, but by helping us outgrow those techs.

- **Rethinking the uses of accumulated knowledge**

  Thanks to the technologies of civilization, humanity has done an amazingly good job of accumulating information in ways that for 99.8 percent of our evolution we could not. Before civilization, accumulated knowledge was limited to what individual elders or storytellers could remember and pass on to their successors. A strong case could be made that the single greatest achievement of civilization has been the escalation of our capacity for accumulating knowledge across generations. By that measure, just to provoke, let me suggest a list of what arguably *might* be the top seven successive inventions of civilization:

  The tablet (carving records in stone)

  The printing press (making multiple copies of a single record—a book)

  The library (multiple books, bringing diversity to human knowledge and thinking)

  The university and its division into academic disciplines (bringing organization of information, the first principle of biological life, to the growing diversity of knowledge and thinking)

  Computer memory (making possible the cumulation of organized knowledge and thinking, in multiple disciplines across multiple generations)

  The Internet and search engines (bringing comprehensive, high-speed access and cross-referencing to the cumulation)

  The tablet or comparable digital-age device (bringing an individual person access to nearly all that has come before, and, if well used, enhanced anticipation of what lies ahead).

By this reckoning, the enormous impact of the internal-combustion engine, which powered our cars and literally drove our economy and refashioned our culture for over a century, would now be outranked by the *search* engine.

Yet, intriguing as that case for the hegemony of accumulated knowledge might be, by the laws of empirical science it is fatally flawed. Are these laws themselves a form of accumulated knowledge? Good question! But at this moment in history, empirical science is the best tool we have. And what it gives us is at least these two cautionary observations:

1. We *already have* a form of accumulated knowledge that is arguably richer than all the libraries and databases on the planet, and has been debugged and refined for five thousand times longer—our DNA. And we have only begun to access it. Hunter-gatherers could only scratch the surface, by learning to pay close attention to body signals with their still evolving consciousness. But now, in the last 0.2 percent of our time, and with accelerating recklessness in the last 0.002 and 0.0002, we have extremely suddenly and severely undermined our DNA's efficacy by disconnecting the bodies and minds it lives in from the environment it evolved in (recall Chapter 3).
2. We've clearly been asking the wrong questions (e.g., how to make more money with tablets), when empirical observation tells us that despite our amazing inventions we're close to failing as long-run survivors.

In short, how can we put any faith in the honored manifestos of our civilization (Bible, Koran, Luther's manifesto, Constitution, etc.) when they purport to provide answers without encouraging further questioning? And when in fact they have a terrible history of being used to *suppress* questioning?

Bottom line: accumulation of knowledge has been in itself an override of evolution to the extent that it gives us crude but powerful intergenerational magnification of our native memories, but it has also, too often, become an anchor chained to the feet of a drowning society. The challenge we have now, then, is to *override that override.* The act of learning needs to be not a passive acceptance of accumulated knowledge as a thing to be built upon, but as a thing to be continuously questioned in every respect, using its own expanding capability to make the questioning more trenchant.

- **Overriding the genetics of impulsive aggression**

It has often seemed to me that wars and genocides are essentially expanded and industrialized versions of the basic bar fight: an aggressive impulse is triggered, and what follows is escalation fueled by emotional vicious cycles, or what academics call "positive feedback loops"—the "positive" being in this case something of a misnomer. The fight spills out onto the street and most of the guys throwing fists or chairs soon retain no recollection of how it started. Or at least that's how it's depicted in movies and TV. I have never personally witnessed a bar fight (just as I've never witnessed a street crime despite living for several years in the middle of what was then the "murder capital" of the nation), so I don't know to what extent the stereotypical bar fight might be one of those phase-3 reinventions of reality brought to us by entertainment or "news" media. But due to the particular circumstances of my life, I have experienced more than my share of hair-triggered violence in other arenas, and what that tells me is that the entertainment media's depiction of a bar fight may be a valid insight about the basic dynamics, if not necessarily the frequency of (mostly male) aggression. A guy thinks his masculinity has been questioned, or another guy is hitting on his girlfriend, or he's being "dissed"—and that's all it takes. *Pow!*

Sometimes the escalation into larger conflict is embarrassingly transparent. President George W. Bush, who avoided military service as a young man, reportedly seized on the opportunity to go to war against Iraq as a belated chance to affirm his alpha-male status. (A former senior official in Bush's administration, Peter Baker, wrote: "The only reason we went into Iraq . . . is we were looking for somebody's ass to kick.") Whether that's what happened or not, I know from what I have witnessed (though not in bars), that small provocations can trigger sudden, violent, reactions.

If it were only a small minority of people with anger-management problems who are vulnerable to being triggered this way, it might not be a big deal for the nation or the world. Those people (Vladimir Putin, Dick Cheney, Wayne LaPierre) could be treated, just as people with broken legs or heart attacks are—and for the normal majority, life would go on as usual. But the problem is that knee-jerk aggression is not just an isolated problem for a few. It's evidently in our genes, as an evolved capacity for reacting to sudden attacks by predators or cornered quarry or adrenalized rivals in a life-or-death struggle for survival. And while the cultures of civilization and domestication have taught most of us to sublimate or resist activating that capacity, our military and intelligence institutions (and sometimes police or prison personnel or football coaches) do the opposite: they train large numbers of young people to become quicker, more unrestrained, and more lethal at such hair-trigger reactions. Marines and Special Forces and SWAT teams are taught *not to think* about social or moral consequences or even the consequences for their own lives, or to ask questions. The pre-civilizational responses are not blunted but honed—and given tech-magnified powers. I find myself wondering: could the unthinkingly thrown fist all too easily become the recklessly launched nuclear missile?

What we do know is that the effects of that training are not limited to the few millions of Marines, soldiers, cops, guards, and paramilitary people who get direct training (and let's not forget the gangs, drug cartels, terrorist states, and others who pick up those same methods); the general public is treated to hundreds of TV shows, movies, and games that make awesome, larger-than-life characters of people who have those skills—from James Bond to Rambo to Jason Bourne to the demure but deadly young former Israeli agent Ziva, in the TV show NCIS. Virtually all of America's kids, teens, and young adults have been treated to such quick-response violence, often with the good guys being the ones inflicting it, and—thanks to the psychological arts of audience identification—have experienced the vicarious satisfaction of such response. It's not surprising that in a sprint culture where stresses of all sorts are on the rise, that kind of arousal too easily spills into people's responses to real-life provocations or irritations. From the time I entered kindergarten until sometime in my 40s, I never once heard of a kid taking a gun into a school and killing other kids, or teachers. When the Newtown, Connecticut elementary school massacre hit the news in 2012, it shocked the country—as had that high-school shooting in Columbine, Colorado thirteen years before. But in the eighteen months following Newtown, there were *seventy-four more* school shootings in America. The disturbances may have festered for years, but now they're erupting everywhere.

- ***Overriding the genetics of addiction***

An addiction is itself an override of biological evolution—a seizing of genetic switches for gratifications that if overused will short-circuit the original survival advantages of those switches with self-destructive effect. In the United States alone, tens of millions of us have become, in multifarious nefarious forms, Frankenstein's monsters. The root causes of addiction are cultural, because it's the drugs, sugar, tobacco, TV, texting, or e-games made available—and pushed on the public by consumer advertising and ever-handy credit cards—that make the short-circuiting easy.

A TV commercial I saw not long ago depicts a boy asking his father if he can use the TV his father is watching, because the other TVs in the house are being used by other family members. The father suggests that the kid go out and play, because "there are limits" (to how many TVs can be used at once in this family), and the boy points out that it's pouring rain outside—a thing the father didn't notice (even though he's right next to the sliding glass door to the outside), because he's watching TV. (He's disconnected, as discussed in Chapter 3.) Whereupon a man who has apparently been listening steps into the house uninvited, and tells the man that with Time-Warner, you can have more than four TVs working at once!

Three things struck me about this ad: First, that the man from Time-Warner had *heard* the father say, "There are limits" (Is Time-Warner monitoring our conversations for the NSA?). Second, that Time-Warner seemed to be suggesting that home invasion is a normal part of its business (the guy just steps in, uninvited), and that it's nothing to be concerned about. And third, that this was the most explicit commercial advocacy for the cowboy economy—and scorn for limits—that I'd seen yet. Why be satisfied with four TVs turned on at once when you can have six or seven? More sit-on-your-butt satisfaction! More GDP! And of course, more millions for Time-Warner and its ad agency.

I'm waiting now for a commercial from Mars or Hershey or Nestle, telling moms that there really need be no limits to the amount of candy or sugar their kids eat. That Time-Warner ad was a particularly shameless message, but only marginally more blatant than the collective effect of omnipresent advertising and marketing of habit-forming candy bars, prescription drugs, fast food, passive entertainment, and other nanny-tech indulgences by our no-limits economy.

Whether scientists will be able to safely cure addictions at their genetic roots is questionable, in part because evolution has been building the survival capacities of the body for millions of years, and the notion of our scientists redesigning one of its basic propensities without unanticipated consequences is hugely problematic. And it's also questionable because modern science is notoriously reductionist and fragmented by its ever narrowing specialization. I wish the neuroengineers well in this endeavor, but I think by far the best chance of overcoming addiction is not to second-guess our brains and glands, but to second-guess the economy of no-limits consumption. If you're a parent with a young kid, your best chance of having the kid avoid the shoals of addiction is to help him or her discover the satisfactions of good food, active outdoor activity, active questioning of authoritarian ideas (including the suggestions of smarmy commercial advertising), light-footprint living, and the long-run satisfactions of reading, hard work, and the search for a meaningful life.

For the Far-Seeing

A strategy for surviving the forces of destruction now tipping the balance of the world must be nimble enough to dodge the thrashing of failing states and their last-ditch attempts to reassert authority, yet steadfast enough to accept that making a truly new beginning for humanity will require overriding some aspects of our nature that have controlled our behavior for millennia—aspects that made us the uniquely inventive and dominating species we are, but that have also ambushed us with a hubris that will finally kill us off if we don't make fundamental changes.

These changes, if successful, won't be like a sci-fi scene in which you are put into a cryogenic box and then wake up a few centuries later completely transformed. They may take generations of sentient experience. And rather than submitting to the still highly risky gamble of genetic engineering, we will most likely have to begin with small changes in how we live from day to day. Among them:

- Consciously assessing the energy source of everything we do, with an understanding that with the exception of nuclear power and its enormous risks, all of our energy (including all our fossil fuels and all our food) is ultimately solar. So, as fossil fuels become too costly or damaging to continue, solar-based technologies are the future.
- Consciously reducing our dependence on technologies that are marketed primarily for purposes of saving effort, because a life well lived is a life of great effort, not great convenience. The more dependent on your techs you are, the more you are controlled by other people.

- Taking note of the extent to which you are controlled by the decisions of people you don't know—at the electric utility, phone company, meat-processing plant, airline, bank, water department, oil industry, Defense Department, Justice Department, NSA, FDA. There are thousands of them, perhaps mostly dependable, but that can change as conditions deteriorate. We can reduce our exposure to broken bureaucracies and rogue companies (remember Enron?) by gradually cutting back on this heavy dependence on strangers. The goal isn't to become more isolated, but to shift from long-distance dependence on anonymous networks to much smaller communities of people you know, or who share common interests and *inter*dependence.
- Keeping alert to resources—both material and human—that may be of value when a collapse is under way. The volume of abandoned structures, broken pavement and concrete, and other discarded materials in the world today is gargantuan, and it will expand as more towns or companies go bankrupt or as the money economy falters and governments lose control. Millions of people in the favelas of Brazil or slumburbs of Cairo, Mumbai, or Kinshasa already subsist by scavenging from the refuse of the wealthy, and in the United States that activity will boom. More about that in Chapter 13.
- Finally, there are those moments in life that precipitate decisions about violence. In my Prologue, I noted that the past century has been the most violent in human history. But it may well be that this is true largely because in the wake of the Industrial Revolution, so much power to control large populations has been assumed by the leaders of nations whose technological capabilities—of mass communication, propaganda, surveillance, and military or police power—have dwarfed those of earlier times. In effect, a few have been able to cause the bloodshed of hundreds of millions—something Attila the Hun or Genghis Khan could not do. But aside from the megalomania of dictators and generals, if you consider people at large, the average human may be less violent now than in earlier historical times, as the Harvard psychologist Stephen Pinkert claims in his book *The Better Angels of Our Nature: Why Violence Has Declined.* People like Gandhi, Martin Luther King, Nelson Mandella and Pope Francis may be the forerunners of a kind of "breeding for peace" that we can consciously practice as we go forward.

12

Breaking Away

Blazing Paths to a New Human Future

An unavoidable implication of the foregoing chapters is this: the free ride will soon be over. For those who have always believed the free ride was mainly about poor people on welfare, or people who seek undeserved entitlements and don't want to work, it should be clear by now that *everyone* has been getting a free ride—the rich even more than the poor, and the big corporations even more than the individual household. All of America, and a growing share of the world at large, at least since the end of World War II, has had a free ride. Now it's coming to an end.

For the rich, the lucrative investments that depended on what was thought to be an inexhaustible supply of cheap labor and free natural capital will begin to fail, as resource scarcity grows and resource wars (over freshwater, fisheries, arable land, rare earths) become more frequent. People who were comfortably cowed when consumer goods were plentiful will grow more restive, and keeping them sedated with bread and circus will become more difficult.

For those among the American poor who have been sustained by generous government support and charity, there will be a painful drying up, as the budgets of state and federal agencies shrink along with the brains of their managers.

For the millions in the shrunken middle, there will be a slowly enveloping and tragic sadness, as it finally registers that the great American myth of perpetual economic growth has been only that—a myth. The disillusionment won't be sudden and excruciating, like the trauma of having a family member die in a school shooting or car crash. But nor will there be any passing of the ordeal. All my references to a "cliff" in the previous chapters are of course metaphorical—and chosen to fit the truncated perception of our 0.002-percent, eyeblink, lives. To bring the image more in line with our real-time sensory experience, when we fall we will fall and fall and fall. . . . The American middle class that has been wired for quick distraction and satisfaction will have no short periods of anger and grief and then "closure." For a civilizational decline there will be no closure.

In contemplating what's coming, I can't help recalling that nightmarish painting by the early Netherlandish artist Hieronymous Bosh, "Visions of the Hereafter"—people tumbling through the final abyss. When I first saw that painting half a century ago, I stared at those wretched bodies falling. Last I looked, they still are. As long as that painting survives, either physically or in some form of memory, those people will fall and fall and fall. It's striking, to me, how similar those Christian visions of humanity's punishment are to the warnings we've gotten from our best secular thinkers and scientists. American politics have caused paralyzing conflict between religion and science, yet both warn of a comeuppance from which, for the guilty or the passive, there's no redemption or happy ending.

Yet, both camps, as well as independent-minded people who may defy convention by having a foot in each camp, suggest that for individuals who are willing to act resolutely and with adequate understanding and commitment of a kind that reaches beyond their own personal devices, there may yet be a future. There are people forming small, conscious communities intent on living in ecological balance with their living planet. It will be a different planet, a far harsher one—the one Bill McKibben calls *Eaarth*—but still perhaps livable under wholly new limits. There are groups of Buddhists, Amish, Quakers, Universalist Unitarians, Makers, and Greens who have quietly begun to contemplate the coming change. The "end times" are still a subject for mockery by the major media, and good material for making blockbuster movies or TV, but for people who can bring themselves to become fully awake, alert, and able to assess the available information without being told by others what to believe, this may truly be a last chance to act.

Doomsday Castle: What *Won't* Work

I have no wish to discourage or disparage the intentions of "preppers"—those people you might run into on hardcore websites or late-night TV who for whatever reasons have concluded that civilization is doomed and that the only hope for their survival is to build a fort. The popular

TV shows "Doomsday Castle" and "Doomsday Bunker" have featured a few of these preppers, as have a ton of books, pamphlets, and websites. There's an old saw that I think applies to these folk: "A little knowledge is a dangerous thing."

I watched a few of these shows and read a few of the prepper books, and noticed that these groups base their expectations on two assumptions that, ironically, may doom their own preparations. One is the persistent belief that the kind of collapse they are expecting will be temporary, and that the goal is to be able to hold out for two or three years of anarchy until order (a different kind of order) is brought about at last.

That expectation, I think, is proof of just how deeply the myth of a perpetually growing economy—one that always recovers from setbacks—has sugared the American mind, even the paranoid American mind. The expectation of just a temporary collapse is a reflexive consequence of having seen that the Great Depression, the Great Recession and other economic crises have always eventually abated and that the march to greater power and wealth (for those who are deserving) will resume. Even for those who hate the American government, there seems to be an abiding faith that after its grip is broken and the regime collapses, replacing it with the *right kind* of regime will bring redemption. But the reality is that while the defense of a well-stocked and equipped refuge might conceivably hope to outlast another passing crisis, a true civilizational collapse is not likely to be reversible within the lifetime of anyone now alive—not even the preppers' youngest kids or grandkids.

The preppers' second mistaken assumption is that the greatest threat to their survival will be the anarchy that arises as the economy fails—the danger of armed looters and marauders roaming the country. I don't doubt that if or when any of the scenarios of collapse described in Chapter 8 are realized, there will indeed be great civil disorder. But the history of catastrophes in our own time suggests that the kinds of civil disorder that result (as in the flows of refugees from war or famine) are more cooperative than predatory, even when people are in the most desperate or dire condition possible. In a world replete with complication and contradiction, it's one of the great *benefits* of domestication: victims of regional disaster or collapse, whether in refugee camps or on their own in storm-wrecked towns, have been historically more likely to work together—and suffer together—in whatever ways they can, than to rob or kill each other.

Of course, there are both cooperative people and predators out there, and always have been. But the preppers underestimate the predominance of the cooperators. And there may be good reason for that. Darwinian "survival of the fittest" became a compelling idea in the century after Darwin's death, and may have strongly influenced such disparate movements as American eugenics, Nazi ideology, and the kind of extreme, anti-government conservativism that abhors giving help to the weak. Maybe it has even influenced me a little, in my apprehensiveness about the amount of effort we make to help people who don't do enough to help themselves. But all of those influences run against the tide of human interdependence. A few decades ago, the mantra of survival of the fittest was found to be an oversimplification and distortion of what has guided human evolution.

When I worked at Worldwatch, I was introduced to the work of Harvard's Edward O. Wilson, and later got to meet him at the Humanity 3000 symposium of the Foundation For the Future in Bellingham, Washington, where a group of prominent scientists had been gathered to discuss the prospects for humanity over the next thousand years. In 1975, Wilson had published his landmark book *Sociobiology*, which suggested that human social behaviors—including the things people do to cooperate, protect, and care for each other—have a biological as well as cultural basis. The book was controversial, and some critics continued to hold that humans are wired to claw each other to death if their survival is threatened. But subsequent research has confirmed Wilson's more nuanced view. A significant finding, for example, is that women are attracted to men who have altruistic tendencies, including a willingness to share resources. What counts in a potential mate is not just being strong or resourceful enough to defend the family, but being socially conscious and far-seeing enough to help strengthen a whole community.

For the doomsday preppers, those two evidently mistaken assumptions—thinking the collapse will be temporary and thinking safety can be secured by a small group armed with guns—converge and collide with one unavoidable reality: to survive requires connecting with more people than are likely to want to hole up together in a bunker, and it requires putting a fair amount of trust in your ability (and theirs) to let cooperation trump a fight to the death. Fighting to the death—over oil, food, water, ideology, or misguided patriotism—is a big part of what has brought us to the brink of civilizational collapse to begin with. So, small battles on the rims of bunkers or ramparts of neo-medieval mini-castles would be merely small convulsions in the throes of a much more sweeping collapse. A PBS documentary about preppers commented, "It's almost irresistible to make fun of them." The narrator noted that a basic premise of the preppers is that "ours is a Mad Max world in which you take from others before they take from you." The narrator then adds: "Yet, trusting in community, not every man for himself, is a hopeful sign for survival."

Civilization led us to attempt violent solutions to the challenges of survival on a tragic scale, but it also brought an irreversible, collective reliance on the hundreds of specialized skills that now keep us alive. That *inter*dependence, along with the cooperation that can make it work, has been in our genes far longer than the inclination to dig in with assault weapons has. If you decide to build a castle—hey, have fun and enjoy the hard work, as I did with my stone cabin all those years ago. But don't deceive yourself into thinking that if you have to retreat there in bad times, you can survive without the help of others. The Doomsday Castle approach is a medieval fantasy.

The Winnowing

Earlier, I referred to the long history of human storytelling as often having been inspired by the undertaking of a journey. Whether in fiction or memoir, a journey of discovery has been a core theme across the centuries and across cultures. It may be both a literal adventure that engages the listener's or reader's imagination as maybe no other form of story does, and a metaphor for the challenges, mysteries, discoveries, loves, triumphs, disappointments, and yearnings that make up a sentient person's lifetime.

A hard reality is that not everyone is fully sentient, and not everyone is up to making a journey that tests the body, mind, and spirit. Maybe everyone was *born* with that capacity, but only a relative few have been fortunate enough to let it mature, the way our abilities to walk or talk mature. Hundreds of millions of others have had their development suppressed, early in life, by circumstances beyond their control—lead-based paint chipping off their walls, or abusive parents, or a surfeit of corn syrup or sugar, or a heavy dependence on nanny-tech supports. In our overpopulated, over-consuming world where hundreds of millions of people are weakened and weakened and weakened (recall the nine signs in Chapter 4), large numbers may not be up to making the hard trek that now lies ahead for those who wish to survive not only as individuals or families but as members of a robust species.

I want to quickly distance myself from several movements that have been galvanized by what may at first seem like similar clarions—most notably the hugely popular "Left Behind" books which propagate a fervent belief that those who have been "saved" will be lifted in Rapture to "meet the Lord in the air," while those who have not will perish in eternal hell. And of course, there are the only slightly less horrific doctrines of mainstream religions, prophesying condemnation of those who haven't accepted salvation. What scientists often say (largely to unhearing ears) is that salvation isn't something you can just accept in the belief that you have been chosen. If you want it, the evidence says you'll have to work for it, and work hard—and connect with a community of others who share your vision. The journey I'm talking about requires gargantuan engagement of body, mind, and spirit in an endeavor that offers no promises of rescues. It will be a daunting trek, and the most daunting part of all may be grasping that *there is no destination*—no "heaven" or "paradise" or "utopia"—only the trek itself. Destinations have sometimes been no more than lollypops to quiet down impatient children, or to distract former children who have never learned how to be patient and envision beyond the moment.

The origin of journey stories may well be the great physical journey prehistoric humans made in the course of their evolution as nomadic, persistence-hunting bipeds who migrated over the earth and occupied it as no other large animal has done. The development of our unique capacities for patient tracking and pursuit, envisioning of outcomes, and great endurance—the traits essential to long-run sustainability—developed hand-in-hand with arduous multigenerational treks over vast grasslands, deep forests, steppes, mountain passes, and eventually oceans and skies.

What's different now is that the physical journey is done (though some of us still seek its satisfactions in endeavors like trail running, hiking, or travel), and in fact is overdone. The traits those millennia of trekking developed, however, are still in our genes. The challenge now—for those who have the will and strength—is to engage those traits in a journey that takes us from our now dangerously dysfunctional civilization to a new form of human future. The essence of it is that we will travel far lighter. I won't be so presumptuous as to say we'll experience things we've never dreamed of, though of course that does keep happening. What I'm fairly sure of is that we'll experience things we *have* dreamed of. That's how humans advance. We learn to envision. What I think I see now is that this will be the most epic journey ever taken.

For anyone who cares to let go of excess consumer comforts and embark on this trek, it's necessary to prepare. Not to be a prepper, but to really prepare. You wouldn't attempt to compete in the Ironman Triathlon, or Western States 100 mile run over the Sierras, without months or years of training. That holds true for the journey of human survival in the twenty-first and twenty-second centuries, as well. For this journey, as for those ritual journeys of endurance sports and wilderness adventures, it must be a focused preparation of body and brain, only more so. For the journey of survival, our best science tells us that a key part of the training will be a strengthening of relationships—between you and your family and friends; between those you know and the strangers you'll ultimately have to put some trust in; between your conflicting tugs of reason and urge; between your own nature as an individual human and the natures of other species you share the planet with; between your present moment and remembered past; and maybe most of all between what you have experienced and what you envision. Without envisioning, the future doesn't exist.

When I say that the physical exploration of the earth that inspired our archetypal journey literature is "done," I mean only that the nomadic spread of our once lightly populated species over the planet is done; all the habitable continents are occupied, and we're no longer lightly populated. I don't mean we're finished with physical movement. In fact, our physiology *demands* movement. But the journey of survival may or may not require moving your home to another part of the country, or to another country altogether. And in any case, that can only be a beginning. What the journey of survival will necessarily require is great movement in how we think, perceive, and envision the road or trail ahead. I'll call this a late phase-2 journey, in reference to the second of the three "phases of disconnection" described in this book's brief early recounting of a civilization that has lost its way.

Phase 1, with its adoption of agriculture and fixed settlements, brought separation from the hunter-gatherer and nomadic life. Phase 2, with the development of more advanced but still utilitarian technologies, enabled movement that could be either physical or perceptual (a car to take you to the Grand Canyon or the movie "Grand Canyon" you can watch on your TV at home). A journey of survival will avoid the more recent phase-3 diversions of redefined reality that have disconnected us from the living world we call earth.

So, this is not about building a refuge and stocking up on food, water, and ammo. Defenders of a fort are basically stuck in one place, just as the civilization they expect to collapse is stuck. The secret of survival is movement, on several levels. Geopolitically, some of those who survive may find themselves having to move around like Geronimo, as global warming causes agricultural zones to migrate to cooler latitudes, or as rising sea level reconfigures the coasts, or as drought and desertification dry up the shrinking Ogallala Aquifer that supplies water to the American plains states. But a phase-2 journey is also a momentous mental and spiritual movement, in sync with the rhythms of nature. It can't work if you panic or rush. It will require many hours of reading, thinking, and talking with others who now grasp what survival must mean. My own ruminations suggest that it will mean recognizing, from the outset, at least the following principles of freedom from the strictures of an outmoded and failing Cowboy/Sprint civilization:

1. ***The path to survival is for individuals and their communities***, not the runaway trains of nations or large corporations. A free society can only be built by still-free individuals who wish to be no longer pacified and weakened by the nanny-tech takeover of our lives and society. It can be built by people who now know that the unanticipated consequence of that takeover is likely to be a widespread deterioration and eventual collapse of the American and global civilizations, which have been fattened on the profits of addictive industries and the looting of natural capital.

   Politicians *will never admit* that such collapse is likely, because if they did they'd never get elected or reelected, or be able to continue for another day waving the banner of an ever better future. When challenged, they will adroitly change the subject or dismiss the challenger as a kook. Most professors in the sciences won't say we're cooked because they'd be accused of trying to prophesy, which scientists know science can't do. (The *World Scientists' Warning*, and subsequent confirmations, have come as close as most scientists dare, with their description of a "collision course.") So, the professors rely on speaking in terms of probabilities. But when we're awakening to the impacts of not one but thousands of factors for any outcome, scientific method can't even yield quantitative probabilities.

   What's left is big-picture thinking, which brings together multiple factors for which probabilities can be only roughly estimated—along with intuitions that are attuned to the movements of nature. As far as we know, artificial intelligence, the great promise of technology, can't do big-picture thinking. Government bureaucracies can't do big-picture thinking. Corporations caught up in the sprint economy are moving far too fast to even try. Only sentient individuals can do it.
2. If enough individuals set out on the hard trek of active questioning, seeking, and learning, ***something akin to a new era of civilization can***

begin. If only a small minority undertake it, we'll likely see a general collapse (the Sixth Scenario of Chapter 8), and that scattered minority can only hope it will endure as a patient protectorate over the span of generations needed for that new era to get under way. The most likely future, as of now: perhaps a few hundred million of the world's people will find ways to become strong, far-sighted, and interdependent survivors. The remaining seven billion or more will find themselves stampeded, one way or another—brought down by those proliferating weaknesses described in Chapter 4. I don't assume that my family and I won't be among those who are brought down. The relatively few who go forward will be those who are able to fully reconsider their current lifestyles, assumptions, and reasons for living.

3. ***A successful survivor won't be either ideologically anti-tech or heavily tech-dependent.*** An undiscriminating anti-tech view is fatal romanticism. The techno-optimistic view—the confidence that technological solutions will be found for whatever problems arise—is its equally crippling opposite. Today's dominant brand of techno-optimism is sheer denial, because (1) it ignores the unintended consequences of powerful technologies, which are often destructive and are bound to worsen as the sprint economy continues to push industries to accelerate obsolescence and increase output, and (2) it ignores the massive extractions from the planet's crust, and resulting pollution and extinction, that have jeopardized the capacity of life on earth to endure.
4. ***This is not about emergency preparedness.*** There's plenty of advice available on how to be prepared for an earthquake, hurricane, wildfire, flood, tornado, tsunami, or major power grid blackout—or terrorist attack, or epidemic. Those events may last hours or days, and a premise of such advice is that sooner or later, normal conditions will be restored. FEMA, the Red Cross, church disaster-relief groups, and volunteers from other parts of the country, and now sometimes even the US military (which is beginning to get the message at last) will rush to help.

   With more far-reaching, civilizational collapse, though, those rescue services may not be available, or will show up with sporadic but diminishing dedication or capability as the weeks go by. The history of regional collapses confirms that likelihood: the day comes when the rescue workers need to return to their own endangered homes or displaced loved ones, and some of the conscripted soldiers shed their uniforms and are gone. If the affected area is large, and certainly if it is global, any restoration of normality will be sporadic at best—and more likely will simply not happen. Emergency supports will disappear. In a true collapse, the long-term probability is that you and your family, or compound or community, will need to be prepared for a very long haul, making a slow and mindful transition to a wholly new kind of life.
5. ***The kinds of people most able to survive a Scenario 6 collapse will be those who know that hard work is more satisfying than passivity.*** I see evidence of that wherever I see people working with their hands

and eyes on tasks they say they'd far rather be doing that way than by operating robotic assembly lines.

In mainstream America, it seems to be widely assumed that artisans and craftspeople are a negligible part of the economy, and that sales of their products are mainly sentimental or whimsical indulgences of the affluent. According to this view, farmers' markets, craft fairs, antique-furnished bed-and-breakfasts, and cottage industries making hand-crafted furniture or pottery or glass, or artisanal cheese or beer, might be destinations for enjoyable weekend getaways, but they're not what drives the economy. As the argument goes, it takes massive factories and office complexes, high-volume factory farms, livestock feedlots the size of airports, coal- and oil-fueled industries, government-monitored mass communications, and cargo ships ten times the size of the Titanic to securely produce the quantities of food, fuel, shelter, equipment, and information needed by the planet's nearly eight billion-going-on nine or ten billion people. You *need* Walmart, Best Buy, Amazon, and McDonalds—and, behind the scenes, Monsanto, Microsoft, Boeing, General Electric, and Exxon-Mobil to make and sell the stuff we use. You *need* Bank of America, Bank of China, NASDAQ, and Fannie Mae to finance and manage it all, and the enormous Departments of Homeland Security and Defense to protect it.

There's enough obvious truth in that view, at least *for the moment*, to make it seem unchallengeable. Without gargantuan quantities of high-fructose corn syrup and bleached-flour junk food, half of the people in East Los Angeles, Detroit, Oakland, or the Bronx would go hungry. Without lung-choking coal (still the dominant fuel for electric power plants), half of our lights would go dark. But as the planet's senior physical, biological, and climate scientists have warned, neither junk food nor junk energy can sustain us much longer. The junk food will kill growing numbers of us, via diabetes, heart disease, and cancer (that "Walk for the Cure" you did will have done more good for you than for the cure), and the coal will go a long way toward killing both us *and* our environment.

The purported remedies we're seeing are far too little, too late—too short-sighted, even by techno-optimists who like to call themselves visionaries. For example, the increasing use of natural gas to replace coal in power plants is still a reliance on fossil fuel, further destabilizing the planet, on borrowed time. So is fracking, which not only prolongs our fatal dependence on a polluted atmosphere, but also contaminates still more of our drinking water—and may be triggering earthquakes.

Free-market advocates may find it inconceivable, but a compelling argument can be made that the kinds of industries now seen as marginal indulgences will need to become the core industries of the human future. OK, yes, we will still need heavy industries at least for a while, to make limited quantities of steel and aluminum, cars and computers. But we won't need them to make many of our houses, a lot of our food, and most of our recreation or entertainment, be it sports or storytelling or music. If affluent people tend to like hand-crafted products, it's quite likely that millions of other people would like them too, if they were affordable. And they'd be more affordable if more people made them, instead of just accepting that we need robotic machines or factory farming to do the job. One of the most clear-headed assessments of American industry I've seen was made over four decades ago by E.F. Schumacher in his book *Small is Beautiful:*

> We need a different kind of technology, a technology with a human face, which, instead of making human hands and brains redundant, helps them to become far more productive than they have ever been before.

I know that prospect might seem hopelessly quixotic, but I recall that when I was a kid in the 1940s and '50s, my mother taught me the virtues of the vegetables she grew without pesticides or chemical fertilizers, which she called "organic"—decades before that word was seen in any chain food store. For the first twenty years of my life, I never heard *anyone* else use that word. Apparently the people over at Rodale Press used it in their pioneering magazine *Organic Farming and Gardening*, and maybe that's where my mother picked it up. But it was as marginal—as below-the-radar of the conventional economy—as could be. I never dreamed, even into my late 30s, that organic foods would become a multi-billion-dollar industry. But it has (US sales of organic foods passed $31 billion in 2012 and are continuing to climb fast), and so have other industries that deliberately refocus on the virtues of hands-on, low-tech, local work or play that gives products a more personal touch but also gives more satisfaction to the workers because they are more physically and creatively involved in the process.

We've been profoundly misled by the assumption that industrial mass production, because it employs technologies invented by higher mental skills, is inherently superior to the work of human hearts and hands. But sadly, for every one person who exercises high-level tech skills to help produce junk food in more efficient or marketable ways, there are a thousand who eat it in ignorance of those skills. And those thousand use less of their mental and manual capacities than do a comparable group who practice artisanal skills. That's because *making* things has been in our blood—and mental wiring—for hundreds of millennia. It would be a cosmic loss to let that blood cool, or that wiring get disconnected, just because we've been persuaded to believe things made by high-productivity industries are more advanced and therefore better.

13

A New Beginning
Essential Steps Toward Long-Run Survival

For those of us who want to survive and take part in the epochal next stage of the human journey, there's a galvanizing question we share: beyond those mindful steps described in the passages titled "For the Far-Seeing" in the first eleven chapters. The Chinese philosopher Laozi famously remarked that a journey of a thousand miles begins with a single step, but ours is a journey that will have no predetermined distance or destination. And let's assume we've already taken more than a few of those first steps of thoughtful preparation and commitment. The question is, given the staggering conditions that are tilting our modern civilization toward dystopia, *what can we do now?*

An immediate conundrum is that letting any established "authority" (political, religious, or academic) answer that for us would more likely lock us in than free us up—if indeed that authority can even acknowledge that we have come to an existential crisis. Those institutions are themselves locked in, as a consequence of their heavy dependence on the technologies that are replacing human self-reliance and community. There's no "how-to" manual—at least none that our top scientists would grant even the smallest chance of providing definitive answers. The path ahead is a wilderness. What follows here is not an authoritarian or authoritative guide, but a set of pragmatic, feet-on-the-ground questions to be considered by anyone who wants to be a part of this epic transition.

Where to Live

To speak of a "journey" toward a rebirthing of civilization can be both metaphorical and literal. With the coming of more severe climate disruption and rising risks of regional eco-collapse, you may need to move physically, if you are now in a place that is too vulnerable to sea-level rise, riverside flooding, drought, disease, or civil disorder. On the other hand, you may have strong roots in the place where you are now, or where you want to be when things go bad. History makes it clear that humans can establish stubborn roots to the places where they grew up—to the extent that they'll willingly endure greater risk by staying than by leaving. It's worth considering, of course, whether those roots were established by family or cultural values or by political indoctrination. (When a soldier five thousand miles from home says he's fighting for his country, did he come up with that assertion himself, and is he really?)

If the decision is to stay where you are, regardless of impending alterations of the place you've known, the journey-in-place may be in some ways harder than if you move to another part of the country or world, as our ancestors repeatedly did for hundreds of millennia. There may be painful tradeoffs: support from trusted neighbors may be greater if you stay, but problems with securing fresh water, food, fuel, and even shelter may be more challenging. To survive in place almost certainly requires giving up most nanny-tech supports and reconnecting with the natural world in ways that we long ago lost in the phase-1 and phase-2 disconnections (Chapter 3).

Escaping the Worst Impacts of Climate Disruption

Scientists can't predict where the next super-storm or thousand-year drought will take place, but general probabilities are well known and can be used to great advantage by prescient individuals. In North America, the risks of personal or family disaster are especially severe in these areas:

- **Atlantic or Gulf coasts** (and, worldwide, all low-lying coastal areas). If I owned a beach house in Ocean City, Maryland, or the Outer Banks of North Carolina, I'd donate it to the Nature Conservancy or Ocean Futures Society (*not* to a beach-loving friend or relative) and get out. Then, for city dwellers, there's what I'll call the Big Apple question: what if you're in New York City or another coastal city where you have employment or a home you can't just "donate" or walk away from? Some parts of some cities may hold out longer than others, thanks to exceptional engineering defenses. If you're a New Yorker and you're lucky, you might not get hit head-on by a category-5 hurricane until you've safely retired or relocated. The seawalls might hold for a few decades, yet—but probably won't. A large consideration is the possibility of forming a citizens' campaign to systematically relocate vulnerable parts of the city and its commerce to higher ground. For New Yorkers, for example, the goal might be to acquire land in New Jersey, Connecticut, or the Catskills for that purpose, rather than wait until a mass exodus is forced. Once an exodus is forced, it's too late for orderly relocation.

- **River Valleys** downstream from large watersheds. Whether in a river city like St. Louis or on farmland along the Mississippi, catastrophic flooding is increasingly likely. People can rebuild (and be able to call on their insurance) only so many times before they—or their insurance companies—are broken. River valleys can be beautiful places, but in an era of heavier precipitation in some regions and quicker snowmelt in others, they are seductively dangerous.
- **Semi-arid or edge-of-desert regions**. Residents of places that have been playing Russian roulette with drought for decades, such as Southern California or Los Vegas or large parts of Arizona, New Mexico, and Texas, may be bucking foolish odds. In 2014, for the first time ever, the supply of water sent from Northern California (Sierra runoff) to Southern California was halted. Areas like the Antelope Valley north of Los Angeles (at the edge of the Mojave Desert) have very questionable futures, although they do have the advantage of becoming major solar and wind-power producers.
- **Tornado Alley**. The science of tornadoes isn't as well understood as that of sea-level rise, but it's a fact that tornado activity has intensified across the country from Oklahoma to Virginia—becoming both more common and more brutal. In the first sixty years of my life, I never heard of a tornado that was a mile wide. In the past few years such monsters have become more frequent, especially in the plains states. Tornado Alley has become Tornado Highway.
- **Ogallala Aquifer lands**. The Ogallala, which stretches from South Dakota to New Mexico and Texas, has been a principal water source for plains-states farms for the past century, but it's a fossil aquifer (containing melted glaciers from the last ice age), not being replenished by rain. And as global warming-driven drought worsens and rain-deprived farms rely more heavily on ground water, the depletion will accelerate. Already, areas of Texas have had to give up farming. Two exacerbating considerations are that (1) the native grassland was long ago wiped out by cattle grazing (the original cowboy economy), causing the land to dry faster and dust storms to proliferate, and (2) a lot of Ogallala land is also tornado-prone. If I had farmland in Oklahoma, I'd donate it to Wes Jackson's The Land Institute, or give it back to the Indians who originally owned it, and get out.
- **One-industry places**. Life is complexity, and in its higher forms depends for its perpetuation on the diversity of its environment. For humans, the most complex of all the life forms we know of, that means a dependence not only on the genetic diversity of the people around us (Hatfields shouldn't have kids with other Hatfields, or McCoys with McCoys), but also on the biodiversity of the other species around and

within us, on which we depend for food, digestion, strong immune systems, and breathable air. But many of the places where people now live have been stripped of their original diversity—reduced to ecological or economic impoverishment. A mining town where everyone's living depends on selling bauxite, or a monoculture town where all the income is from Monsanto corn, is not a promising place to be when TSHTF. We'll need companions or neighbors with diverse knowledge and skills, and we'll also need places to live that have a (relatively) wide variety of remaining resources to draw from—and to provide resilience should we draw too much.

Where to Go, Then?

If a principal determinant of habitability will be climate change, the map of high-risk areas can be complemented by a more encouraging map, of more promising criteria:

- **Higher-elevation** places have two advantages: they may be cooler, at a time when some formerly habitable regions are becoming intolerably warmer; and they may be better for growing food, as agricultural zones move to higher elevations and latitudes. And of course, most of these areas are not vulnerable to coastal super-storms or flooding.
- **Higher latitude** can offer the same food-growing advantages on a warming planet as higher elevation. (Conversely, some formerly temperate areas may become newly amenable to growing subtropical or tropical crops.) There's also the potentially troubling fact that not only agricultural zones, but the ranges of mosquitoes, killer bees, and other pests may shift farther from the equator as the planet warms and the region of freeze-free winters expands. Population density may decrease (reversing the recent trend) in some southern states, but increase in the northern-tier states from Washington and Idaho across to New Hampshire and Maine. In the 1970s, young men who objected on religious or moral grounds to killing other young men in Vietnam sometimes considered moving to Canada. With global warming, we may see whole families and communities going north. (Along with leaving too-hot places behind, emigrants may find areas of Canada that were once too cold turning more temperate.) Canada's powers-that-be may not like the influx, but may be pleased with the expansion of their nation's temperate land-mass, at least as long as their nation remains a nation. In the southern hemisphere, of course, the polarity is reversed and higher latitudes are farther to the south. Patagonia may attract some of the same kinds of dystopian refugees as Canada.
- **Repurposed places.** Over the past century, the world's population has become steadily more urbanized—people abandoning small towns and farms and moving to the cities and suburbs. America is now dotted with places where people have left and property can be

had cheap. For people seeking a new path and a life that subordinates tech services to human strength and is closer to the land than the life of a city street or airport lounge is, these places have potential. As the realities of incipient dystopia begin to register more widely, the rural-to-urban migration may for small numbers of people reverse. For a few, that began in the 1970s, but with the knowledge we have now there will likely be a quiet resurgence. The repurposing of barns, silos, firehouses, schools, and small-town churches that have lost their congregations is a growing trend. Here and there, individual families or small communities have acquired rural properties and brought them back to life—literally. A cautionary note, here: in a world that is heavily dependent on long-distance transport (of food, fuel, and people themselves), dispersing most of the population would be a bad idea. There's energy efficiency in high-density living. But that efficiency is nullified if the food for New York is transported thousands of miles from Chile or Spain, or the energy from Texas or Saudi Arabia. On the other hand, if the economics of food and energy become localized, dispersal into small, self-sufficient communities is healthy—and safer than waiting for breakdowns of urban society. America now has a huge supply of old buildings and infrastructure that can be repurposed. A former firehouse that's not capacious enough for housing high-tech firefighting equipment might be ideal for a small, low-tech business. A nineteenth-century Pennsylvania barn can be turned into a high-ceilinged, skylighted, solar-powered house that will outlast a house built two centuries later—and might even last into the genesis of a new civilization.

- **Diversity Hotspots.** A supplement to the caution about "one-industry places" in the previous pages is that in one respect, at least, urban areas offer a distinct advantage—especially small cities in which a wide variety of industries and cultures have coexisted peacefully. Of course, wherever you go, there are tradeoffs. But overall, one of the best choices might be a culturally diverse, high-elevation, high-latitude college or university town in a biologically diverse, virgin-forested region surrounded by rich soil, high mountains, and a large wilderness watershed where no gas stations or heavy industries have ever operated. If you can find a place like that, please let me know.

What Work Can You Do?

When our civilization is staggering but not yet broken, a lot of occupations that are now marginal will become more central to survival—and critical to our being able to visualize what will be possible later. More people will study, document, and practice preventive health care by means of optimal nutrition, physical and mental exercise, stress reduction, and elimination of environmental pollution (including pesticides and herbicides, processed-food additives, and genetically modified foods), and fewer will continue to rely mainly on conventional medical and pharmaceutical industries. It is vastly less expensive and less tech-dependent to prevent illness than to treat it. More people will be mental-health practitioners focusing on relieving environmental and social stresses, while a lot of psychopharmacologists will probably go the way of psychoanalysts. More people will work as teachers or rangers taking kids on field trips to see wildlife or to experience toy-free play, while fewer will work at selling toys or electronic nannies.

As funding retreats from services once taken for granted, one of the great losses will be libraries. It may seem counterintuitive, but the hopes we have for the readers and thinkers of a century from now may never be realized unless far-seeing people who are not librarians or curators take certain critical steps very soon. One of those steps, both momentous and mundane, is to develop a simple method of archiving the best of our civilization's knowledge without depending on digital technologies that become quickly obsolescent and that depend on fossil-generated electric power. While the libraries rush to convert books and other printed materials to digital records that cost fortunes to implement, they may be racing toward a technological train-wreck. What far-seeing groups need now is a practical system of digital archiving that will not be scuttled by either the rapid obsolescence of complex systems or the coming power blackouts—or the defunding of libraries altogether. For a sentient community with a list of a few thousand cherished books, articles, maps, and videos that represent its collective choice of what legacy to pass on to its great-grandchildren, I'd guess that something as simple as an ample supply of e-books and solar-powered rechargers will provide the portability and practicality needed for the coming years. Free-standing solar rechargers, both for e-books and for electric vehicles, may be among the most important items to acquire or assemble before it's too late to find them.

What matters is that the protectorate communities will have begun the trek with at least some ready productive and educational capacity. An early start—while the lights are still on—enables each group to develop a fairly self-sufficient micro-economy: conserving water, producing food (supplemented by stored supplies for several years), shelter-building (especially from salvaged materials), and, ASAP, some trading with other groups, for food or other resources. And, meanwhile, never losing touch with the best and worst thinking of the past five thousand years. A key to outwitting dystopia is having a strategy worked out well before the SHTF.

- **Later in the transition**: Dystopian or not, the economy of the future will be profoundly different from that of today—and in very different ways than I think most of today's investors or masters of the universe expect. A lot of the now confidently presumed paths to success will lead to dead ends. In response to the heavy ecological hoof-print of the cowboy economy, new light-footprint occupations that barely register in public awareness now will go mainstream: alternative energy technology, rainwater harvesting, and small-scale organic agriculture, or permaculture, will push aside petrochemicals, central power plants, and coal. A substantial volume of car use will give way to bicycle use and walking, as more people give up long commutes and gravitate toward more localized communities and economies—or as the reliability of motor vehicle manufacturing, sales, and service breaks down and as the fossil-fuel economy begins to die. Suburbs may decline, while cities reverse their sprawl and small towns see some repurposing and revival. In response to the ecological stress of the sprint economy (which produces unnatural physical and psychological stress in humans), a culture of conscious slowing will create new jobs of a kind that may seem exotic today but will become more common as the present rush proves unsustainable. "Slow food" will slowly catch up with fast food, in a classic tortoise-and-hare (or persistence-hunter) fashion. Eventually, the overextended, ecologically destructive industries of burgers and fries will collapse. More people will teach subsistence cooking, meditation, or endurance sports, while fewer will grow up (or just get older without growing up) to be excitable sportscasters or bullying corporate cowboys. Fewer commentators will glibly criticize people who seem ambivalent for not "having the balls" to do something decisive or disruptive, while more will discuss *why* a person might be wisely hesitant or cautious. More people will sell free-range eggs, chicken manure, pecans, and chard at local farmers' markets, while fewer will be cashiers at Walmart. More will make and sell spades and rakes, and fewer will make or sell leaf blowers or chain saws. More will practice neuropsychology, while fewer will do plastic surgery. Many more will sort through the remains of the dying consumer economy looking for junk they can sell, barter, repair, or recycle.

Who Will Work with You?

You can't be fully alive (or alive for long) during the collapse if you're a recluse. You'll need to find good working relationships with others. There may be loners holed up in mountain cabins, with their shotguns and cans of beans and no friends, but if the United States becomes a failed state, once a guy's beans are gone or he gets sick or injured, or his cabin gets flooded or invaded by black widows and black mold, or his pickup truck's axle breaks or his gasoline runs out or his horse dies or he runs out of feed for the chickens or he gets bitten by a rattlesnake or begins to hallucinate—he's done.

As the biology of sex has taught us, there's no future for individuals if there's no future for human relationships—for people who love and care for each-other and their kids. And as the biology of evolution has taught us, it takes biodiversity as well as fertility for a community or species to thrive. And as the troubled history of technological civilization has made clear, a viable future depends not only on a diversity of genetic talents but on a diversity of learned skills. In practical terms, that means living with a group of people who have a variety of special skills in addition to having rediscovered and cultivated their evolutionary heritage as good generalists. In my reflections about the time to come, I've brainstormed a list of the people I'd like to have go with me on the journey—either as neighbors if I base my future life physically where I live now, or as an assembled community of kindred souls if we decide to move. Your list might be different than mine, but scan this and you'll get the general idea: to go forward toward a world of renewed life, you also need to be well grounded in the basics of physical and emotional survival. In my cogitation, I imagine that some close friends and I decide to buy an old farm (with old but well-built barn and silo, well, and a manageable piece of land) and form an enlightened community. Who would I want to be with my family and me?

- An organic gardener with a love of life and a big straw hat
- A great cook
- A plumber/water-systems expert
- A carpenter/structural engineer
- A solar power technician/electrician
- An ecologist
- A climate scientist
- An expert on wilderness survival and forage
- An historian with a solar Kindle full of good books
- A book lover with a library of good books, movies, and music
- Several school teachers, both male and female
- Kids of all ages
- A young doctor/endurance athlete
- A neuroscientist/psychologist
- An evolutionary biologist

- A good pianist or guitar player who likes classic rock
- An artist who likes to work with found objects or broken glass
- A writer with a sense of humor, to keep records of our journey

I would like everyone on the list, in addition to being a competent specialist in his profession, to be committed to getting well educated in related fields. I want the doctor to be competent in conventional medicine, but also in holistic health and nutrition. I'd like the organic gardener to have strong knowledge of agricultural history and ecology. I'd like each adult to mentor a kid.

But the most important thing to recognize about this list is that you'll need to make your own.

Security

In an earlier passage I was briefly dismissive of the Doomsday Castle approach—and said no more at that point, for a reason: if you get caught up in fear of attacks by marauders or, worse, in fights over resources or territory, you will have failed in your mission despite your dedication. A brutal possibility is that viable small communities of the far-seeing will be like the scattered acorns that eventually produce young oaks. It may take hundreds of acorns to produce one tree, after the squirrels and forest fires and decomposition have done. The odds may be better for a shepherd of the new humanity than for an acorn, but it should be understood that if you succumb to bunker mentality, you've already become another victim of the collapse.

That doesn't mean that to succeed, you have to be a pacifist. I have wrestled with that question for half a century, having been born during World War II and grown up in a Quaker family. I know now that a certain degree of defensive preparation is essential, but it can't become an obsession, or you're lost. For optimal security in a time of civil disorder, probably the best protection will be our very below-the-radar, inconspicuous but close-knit communities—small clusters of fellow survivors who are quietly looking out for each other.

Sidestepping Civil Disorder

What we face is unlikely to be just a Mad Max world of sociopathic men riding monster trucks or motorcycles and looting the derelict wreckage of a suddenly destroyed civilization. In most of the scenarios described in Chapter 8 (with nuclear holocaust being the stark exception), the collapse won't happen in a day, and the efforts of shocked survivors to restore some kind of order are likely to be concurrent with the gathering collapse. Moreover, there are likely to be sporadic regional breakdowns (beyond those already occurring now) in advance of a more sweeping collapse: more frequent power grid blackouts, more drought-driven crop failures and wildfires; more busted firefighting budgets; more pipeline explosions, stock market shocks, and empty reservoirs. And, soon, more Internet failures, more unfunded government programs, and municipal bankruptcies—and more widespread violations of human cooperation and trust.

Beyond security (and food-security) concerns, there will be the inevitable efforts of hundreds of scattered or stranded groups to form provisional governments and restorations of civil order. Some if those efforts may constitute as much of a threat to our long-run quest as do the marauders in the monster trucks, because the default impulses of people who didn't think ahead will be to try to resurrect the kinds if authoritarian, free-market, and nanny-tech regimes that brought the fall. Our protectorate groups will need to act with great discretion, especially in how we educate our children, in order not to provoke interference by hastily assembled groups with other plans and discredited kinds of schooling.

Unanswerable Questions

We can speculate endlessly about what the course of collapse will be—whether it will be triggered by one of the scenarios listed in Chapter 8 or by a toxic mix of them, or by a cascade of new troubles triggered by a fatal tech-dependence. Whatever the precipitating cause, by the time the collapse is full-blown, hundreds of millions of people will likely have perished—whether in civil wars, water wars, rare-metals wars, starvation, epidemics, genocide, a metastasizing of the kind of hatred embodied by ISIS or Al Qaida, or coastal mega-cities destroyed by super-storm surges. So, as we go forward with our quiet trek, there will undoubtedly be moments when we feel the urge to look back over our shoulders and wonder: what is happening to those who are *still* in denial, or who are too weakened or numbed (recall Chapters 3 and 4) to venture away from what is at least familiar to them?

We will have a thousand questions, but very few clear answers. As nations fail, what will become of their military arsenals and troops? Will their abandoned cities seize control of whatever is within their reach, including military bases? As the cities, too, fail, will opportunistic militias seize territory and resources? Will the dying world we've left behind enter a new Dark Age of cutthroat fiefdoms, in which people who were once lawyers, industrial designers, or teachers have become serfs or slaves? What will become of the growing masses of sick, disabled, or mentally ill people for whom services have ceased? But for us, those questions will all be unanswerable. Even future historians, if there *are* any, may never have the full story of what happened.

One thing we can envision with some confidence is that when the collapse is full-blown, and denial is no longer possible for most, there will be many remaining survivors—perhaps a majority—for whom the "better angels of our nature" (as Abraham Lincoln put it) will prevail, and who will mount their own belated efforts to save the day. Not everyone out there will be marauders. But at that late stage of awakening, it will almost certainly be too late. Those belated efforts will not have the advantages *we* have, of working within the still largely intact framework of a society where money still has value, police are still on the job, hospitals are still staffed. And there's another disadvantage, too, for those who come around too late: in a civilization that has taught us all to be specialists, it has been natural to specialize in our public-interest efforts, as well. We have groups that specialize in rural search-and-rescue, urban emergency medicine, firefighting, food for the hungry, foster-care for children, and so on. But the communities that form the protectorate will need to be just the opposite: deliberately diversified in skills and imaginations. Those who take longest to recognize the human plight are likely to be the ones who have least exercised their ability to envision, and may be least able to know what on earth to do at that late hour.

That urge to look back may be irresistible, because those who embark on the journey early will be self-selectively strong in both curiosity and empathy for our fellow humans. But I'll suggest a useful analogy here. When airline personnel give their routine briefings to passengers on what to do in case of the need for oxygen masks, the advice for parents accompanying children is to put the masks on *themselves first*. That way, if they have only seconds to act, they have the best chance of being conscious long enough to save both themselves and their children. For us, resisting the urge to look back at our crippled world—and quite possibly to get distracted or sidetracked—is like donning that oxygen mask on a crippled plane. We can think of the children awaiting our help not only as our own kids but as the descendants we hope to have for the next ten thousand years.

When Do You Make Your Move?

The difficult but critical first step, in preparing for an unthinkable yet nonetheless very likely disintegration of our world, is recognizing that the breakdown has already begun. For millions, that first hard step—that simple recognition—will be *too* difficult and will never be taken. Denial is pervasive, but even when the course of events is undeniable, the patterns of past behavior suggest that for many it will persist until too late. And beyond denial, there are all those questions that paralyze initiative: How can I leave my house, if the real-estate market has crashed and I can't sell it? How can I leave my job, if no one is hiring? How can I just cut all the ties I have to the life I'm living right now?

It's a true dilemma, because those questions are legitimate and there are no easy answers—except, perhaps, that if you don't cut the ties as part of a plan over which you have some control, the coming disorder will cut them anyway. There's nothing to be gained by pondering whether the dystopian prospect is far-fetched, and there's everything to be lost by delay. And for anyone who still needs a short summary, take one hard last look at the evidence that we live in a culture beset by widespread obliviousness—and by forces that promote obliviousness for political or ideological purposes. Large majorities of our fellow humans still do not know that we're in the throes of epochal climate disruption coinciding with a population-growth spike that has added more people to our finite space in the past 0.06 percent of human evolution than in the previous 99.94 percent. The global population reached 2.5 billion over the 24,000 centuries of human evolution before I was born, then more than tripled in the less than one century since. The earth, on which we depend not only for all our food, water, and air but all our personal space, privacy, and peace, hasn't grown at all. And as for the technologies we've been told will solve all problems, their overall effect has been to make us more dependent and weaker.

So, the time to make the move, or at least begin active preparations, is now.

Beyond Survival, What?

This is a mystery, because the human journey has always been a journey of discovery, especially when you don't really know where you're going. (You may know where you're going *geographically*, but you don't know where that will take you in terms of spirit, understanding, and perspective—and capacity to be part of an epochal transition that reaches far beyond your own moment.)

The honest answer to that question of where we're headed beyond our own survival is that *finding* the answer will be a primary task of those who make the journey, and their children. I'm fairly sure, however—thanks to what we have learned from the science of human evolution—that the answer won't come to us "out of the blue," like a revelation to Moses or the young Buddha, or Jesus—or, later, by the rush of technological invention that came with the light-bulb breakthroughs of Edison, the Wright Brothers, Einstein, and Bill Gates. Experts on innovation now seem to agree that those revelations of individual genius are nearly done, because most breakthroughs today are the products of collaboration and testing by many researchers, and approvals by many officials. Genius brought us great miracles of human triumph in a Darwinian struggle, but has also baptized us in a sea of new powers that have ironically rendered us increasingly numbed and weakened.

For the minority of us who can escape that immersion and get our heads above the water enough to survive the moment and envision what might come next, the key to moving ahead is better understanding what we've already been through. That understanding can be found in our DNA, in our history and prehistory, in the annals of our sciences, and in the biographies of our most visionary individuals and our most sociopathic or narcissistic ones. It's in the literature of human passions, quests, discoveries, and invention—and hubris and blindness, and tragedy. So, for our journey it will be important to bring along that solar-powered home library-on-a-Kindle (see the Appendix of this book, "Further Reading and Resources").

The daily work of the mid-to-late twenty-first century survivors, then, will not find us foundering in the Network of Things (*NOT*). More of our work will be outdoors, reconnecting with the earth and progressing toward a mutual regeneration—restoring habitat and rediscovering our own nature. And along with practicing the physical arts of light-footprint home-building, a big part of the daily work will be *education*—of ourselves and our children together. The curriculum will need to emphasize meshing with the rhythms and cycles of the natural world, which in practical terms will mean slow food, artisanal industry, patient reflection on what we discover or learn; and the celebration of hard physical work and endurance.

The curriculum will also need to include exhaustive cultural forensics, to illuminate what went wrong with the sprint economy, and how on earth our twentieth- and twenty-first-century economists could have so unquestioningly believed in—and promoted—a doctrine of perpetual economic growth. It will need to emphasize the long-run survival of humanity, and a new economics that does not "discount" the value of future generations, as current economists routinely do.

Perhaps above all, this curriculum of the protectorate will value its children as no previous generations have: not "spoiling" them or giving them notions of entitlement, but rather recognizing the clarity of vision and wisdom that becomes evident in kids within their first two or three years—a clarity that too often is first regarded by adults as "cute" or "endearing" but is soon subdued as excessively hyperactive. In the new education, we won't train our kids to be like us. We'll protect them from harm as we and they learn together how to protect *their* kids and grandkids—the future of humanity.

The Sum of All Hopes

In the rhetoric of twentieth- and twenty-first-century American politics, "hope" is a palliative—a reassurance to troubled people that with strong leadership and faith, there will be great rewards ahead. In the land of the free, the prevailing belief seems to be that hope is free—you need only embrace it, with your vote or your church attendance.

But real hope isn't free, just as real freedom isn't free. Both are achieved through hard thinking, envisioning, and work—and maybe, too, by a hunger for real adventure. Genuine hope and freedom can never be truly guaranteed. They can't even be honestly promised, say nothing of *given*, because no religious authority or government institution has ever had the power to fulfill such a promise.

Unlike Heaven or Paradise or the Good Life on Earth, which are all routinely promised by a few to the many who are passively compliant, real hope is akin to what might happen in the blood and nerves of an astronomer who has suddenly discovered a dauntingly far-off but evidently life-supporting planet, after a lifetime of diligent searching.

The planet we are searching for is our own Earth. We are the descendants of a hundred thousand generations of successful survivors, persistence hunters, and migrating explorers—descendants of the quick-witted few who did not get killed by predators, snakes, disease, falls from cliffs, or fights with rivals before they could reproduce and raise children, who in turn did not killed off.

The strengths of successive species of humans throughout those hundred thousand generations of survival have been patience, endurance, and vision—not speed, not power, not even a high level of inventiveness. From the beginning, the probability of *our* being here now must have been almost infinitesimally small, like the probability of tossing a penny and coming up with a heads a hundred thousand times in a row. (For the religious, what a gift is *that!*)

Yet, after all that, our bloated population is about to blow it, and turn tail. Somehow, in just the last tiny fraction of 1 percent of that miraculous run of success, the few who control the fate of the many have largely abandoned those qualities of patience, endurance, and vision in favor of a sprint economy that promises quick rewards, quick returns on investment, and ever more rapid development of the technologies of distraction and consumption—which in turn are killing off our real survival skills, and for many, all hope.

The big promises—of the American Dream, the Christian Salvation, the Muslim Paradise, the Pepsi Good Life—have all proved empty. What remains now is hope—not the empty rhetoric of freely dispensed political clichés, but the great strength that comes with real vision and perseverance. Those of us who hope to be part of a protectorate that will let us reconnect with our real nature—and survive the aberrations of the industrialized soul—do not have a plan, because now we are embarking on a journey to new territory. We don't have a destination, because even far-seeing people have only been able to see a few steps of the actual ground ahead. Right now, we just need to keep moving.

Just a few centuries or even decades ago, there appear to have been only a few people who were still strongly in touch with those ancient human strengths that industrial civilization has eroded. People like St. Francis, Buddha, Thoreau, Gandhi, Confucius—and the reclusive Helen and Scott Nearing of Maine, perhaps. Most, like the Nearings, probably kept low profiles—as *we* will need to do—because they were perceived as threats by the dominant institutions. They probably numbered, worldwide, in the thousands.

Now, *thanks* to the vastly increased digital connectivity of our civilization (can we appreciate the irony?), there are undoubtedly millions of people on this earth who understand our nature—our *humanity*—in the ways those enlightened individuals of the past did. And now, too, many keep a very low profile, not necessarily for fear of persecution (although that is still risk), but because they've recognized that big media news, entertainment, advertising, and publicity are largely controlled by the dominant institutions that have put us on what our best scientists have called a "collision course." Prudent and quick-witted people are quietly stepping off that course.

We can't know our real numbers, but all indications are that our numbers are growing. Our millions are still only tiny fractions of the billions of humans now occupying the planet, but they are enough to give us great hope—real, worked-for, hope—that we can form the number of independent and conscious communities we will need in order to move forward. The challenge is to use our vision and our wits, not to move directly against, but quietly over, around, and right through the impediments of a civilization that has grown too hubristically, too violently, too fast, and too disconnected from its origins. We have everything we need, at least to get started, right now.

Appendix: Further Reading and Resources

Who will survive and thrive is always to some degree a matter of luck, or *chance.* But as the medical pioneer Louis Pasteur (whose discoveries enabled many who came after him to survive) observed, "*Chance favors only the prepared mind.*" The purpose of this Appendix, if you have read this book and are persuaded, is to improve your family's or community's chances of making it through the coming catastrophes. You can't predict when the next hurricane will strike where you live, but if you live on a vulnerable coast you can choose to move before it happens. Maps showing relative risks are available. The following pages provide information on those maps and other potentially life-preserving information in the following eighteen areas of dystopian concern:

- Water Outside the Coming Water Wars
- Food Without Industrial Agriculture
- Health Without Corporate Medicine
- Refuge from Climate and Extreme Weather Catastrophe
- Refuge from Civil Disorder and War
- Shelter in-place and on the move
- Off-Grid Energy
- Off-Web Communications
- Light-Footprint Mobility
- Light-Footprint Spirituality
- Safekeeping of Knowledge
- Freedom from Excessive Tech-Dependence or Addiction
- Conscious Communities . . . and Country without Empire
- Post-Global Economy
- Envisioning the Future
- Enlightened Education
- Emergency Preparedness
- The Warnings of Science

Note: These listings are far from comprehensive; they are a sampling selected for diversity of information and perspective, and most will provide links to additional sources. A listing here does not constitute an endorsement by the author, who encourages considering a wide range of perspectives but whose own views will in some cases differ.

Water Outside the Coming Water Wars

Drought area is expanding in the United States and worldwide. But even in areas with ample water for now, the deterioration of water infrastructure is a growing liability. Along with oxygen, water is the most essential survival resource, and no matter what shape-shifting occurs in human governance and industry, everyone—no exceptions—will need better preparation for the supply, security, and purity of fresh water than most of us have now.

- Basics on wells for home owners, http://www.wellowner.org.
- Rainwater harvesting, http://www.theecologycenter.org.
- Global Water Policy Project, http://www.globalwaterpolicy.org.
- Climate-change impacts on water supplies, http://www.climatehotmap.org/global-warming-effects/water-supply.html.
- Owen, David. "Where the River Runs Dry: The Colorado and America's Water Crisis," *The New Yorker*, May 25, 2015.

> *"Water is the driving force of all nature."*
> –Leonardo da Vinci

Food Without Industrial Agriculture

As global population has spiked, human dependence on industrial mass production of food has also spiked. That dependence has brought a panoply of ills: monoculture that is eroding biodiversity; chemical fertilizers and pesticides killing marine life (via runoff) and evidently killing humans (via cancer); and the damaging health impacts of commercial food products containing hundreds of chemicals—some of which have been implicated in cancer, heart disease, diabetes, and other causes of human weakening, and most of which have never been tested for long-term safety or health impacts. Chances of survival are greatly increased for those who disengage from dependence on these industries—most of which are not designed for health benefits but for the profits of their managers and investors.

- "A Plan for Food Self-Sufficiency, www.motherearthnews.com/homesteading-and-livestock/self-reliance/food-self-sufficiency.

- Permaculture: https//en.wikipedia.org/wiki/Permaculture
- "How to Grow Organic Food for Family Economic Survival," www.verdant.net/food.htm.
- Shiva, Vandana. "Eat Local: Rebuilding the Broken Food System" (India), *Countercurrents*, July 16, 2015, http://www.countercurrents.org/shiva/160715.htm.
- Schlosser, Eric. *Fast Food Nation: The Dark Side of the All-American Meal*. New York: Mariner Books, 2012.
- The Self-Sufficient Homestead: Surviving Civilization on the Homestead, Audio Podcasts, http://sshomestead.com.
- Ayres, Ed. "Will We Eat Meat?" *Time* Magazine, "Beyond 2000, Our Planet, Your Health," November 8, 1999.
- Food Tank, http://www.foodtank.org.
- Thayer, Samuel. *The Forager's Harvest: A Guide to Identifying, Harvesting, and Preparing Edible Wild Plants*: Forager's Harvest Press, 2006.
- Restorative Food Systems, http://www.bioneers.org/restorative-food-systems.

> *"If you can't feed a hundred people, then just feed one."*
> –Mother Theresa

Health Without Corporate Medicine

Prevailing views of health in the United States—as shaped by mainstream politics, media, and business—are as blind to the fundamental failures of health care today as they were to racial injustice a century ago. Never mind such shams as the 5K you can run (for a fee) to support cancer research, when we already have knowledge which if widely applied would reduce the incidence of cancer, heart disease, and most other diseases by as much as 90 percent if that knowledge were not being ignored or even suppressed by the dominant institutions. A key strategy for long-run survivors, now, will be to implement that 90 percent reduction for their own families or communities, via serious preventive medicine above all else.

- *Common Ground* magazine, http://commonground.ca.
- National Center for Complementary and Integrative Health. https://nccih.nih.gov.
- *Preventive Medicine* Journal: http://www.journals.elsevier.com/preventive-medicine.
- American College of Preventive Medicine: http://www.acpm.org.
- "Survival Benefit from Leisure-Time Running" Antonio Benicio Fernandez, M.D., FACC, American College of Cardiology, January 6, 2015, http://www.acc.org.

- Bramble, Dennis and Lieberman, Daniel. "How Running Made Us Human." *Nature* 432, no. 7015 (November 18, 2004): 345–52.

> *"Without health, life is not life."*
> –Buddha

Refuge from Climate and Extreme-Weather Catastrophe

No one can avoid at least some of the damaging impacts of escalating climate change—such as the shrinking of arable land area due to expanding drought and desertification and its effects on food production and prices (and on profits for investors in agricultural industries). But smart preparation can greatly increase the chances of avoiding the most devastating impacts, such as sudden obliteration of homes by hurricanes, floods, or tornadoes, or the cascading collapses of regional economies and the systems (water, power, communications, banking, government) that provide employment and sustenance for many of us.

- Aspinall, Richard John, ed. *Geography of Climate Change.* London: Routledge, 2012.
- Climate Hot Map: Global Warming Effects around the World, Union of Concerned Scientists, www.climatehotmap.org.
- Maps of climate-change impacts on the US, National Climate Assessment, www.nca2014.globalchange.gov.
- Mapping the Impacts of Climate Change: Extreme Weather, Sea Level Rise, Agricultural Productivity, and Overall: Center for Global Development, http://www.cgdev.org/page/mapping-impacts-climate-change.
- United Nations Framework Convention on Climate Change: Freshwater, Food, Human Health, Forests, Coastal Populations, http://unfcc.intg/essential_background/the_science/climate-change-impacts-map.
- http://www.worldchanging.com.
- National Climate Assessment, US Global Change Research Program, http://www.globalchang.gov.
- McKibben, Bill. *Eaarth: Making a Life on a Tough New Planet.* New York: St. Martin's Griffin, 2011.

> *"North America's Appalachian Mountains may be a safe haven from climate change, according to a new study by the Nature Conservancy."*
> –Huffington Post

Refuge from Civil Disorder and War

Survivalist blogs, newsletters, and magazines—and books like *Patriots: a Novel of Survival in the Coming Collapse,* by John Wesley Rawles—have put much emphasis on scenarios of looters, marauders, and desperate refugees pouring out of the cities (or into the country) when "it" happens. Websites bearing the acronym WTSHTF have proliferated, offering advice (and ads for equipment and supplies) about the best guns, dried foods, and "bug-out" vehicles. A few of those sources are listed here. But the best strategy for long-term refuge may be to find a place that is far from the expected disorder (and therefore very hard to reach when the fuel runs out, or when soft people get tired), and where the best defense is not guns and bunkers, so much as a tightly knit community equipped with skills, knowledge, and long-range vision, *in addition* to reasonable defensive capability.

- Collapse of Industrial Civilization, http://collapseofindustrialcivilization.com.
- Off Grid Survival, http://offgridsurvival.com.
- Survival Life, http://survivallife.com..
- Doomstead Diner, http://www.doomsteaddiner.net.
- Survival Magazine, http://www.facebook.com/SurvivalMagazine.
- Preparing for SHTF, http://prepforshtf.com.
- Armageddon Online, http://www.armageddononline.com.
- Isaac, Jeffrey. *The Outward Bound Wilderness First-Aid Handbook*, 4th ed. Falcon Guides, 2013.
- "What to Do during Civil Unrest in Your City," http://prepforshtf.com/civil-unrest-city.

> *"According to researchers at Cornell University, . . . if you want to improve your chances of survival you should head for the hills."*
> –SHTFplan.com

Shelter, in Place or on the Move

The meaning of shelter may shift in the coming days, depending on how rapidly or slowly the deterioration of civilization proceeds—and depending on how early or late you make your move. For those of us who get started soon, while the money economy and government agencies are still functioning, there's a chance to make full use of long-run survival technology: off-grid alternative energy, fresh water harvesting, subsistence food production and storage, and structures located and built to withstand extreme-weather events.

For those who get mobilized later, when systems are destabilized or failing, it may be necessary to forego any thoughts of permanence, at least for the time being, since one of the requisites for a journey of survival, once the collapse is underway, is to be adaptable and resilient—more like the Native Americans who could pull up camp and disappear, than like the idealists of ancient Athens or Renaissance Florence, who were confident they could build for the ages. The Athenian and Roman regimes are long gone, and now modern Greece and Italy, too, are in trouble.

- Hurricane-resistant structures: Deltec Homes, 69 Bingham Road, Asheville, NC, 28806, http://www.deltechomes.com.
- 50 Machines: Open Source Ecology: https://en.wikipedia.org/wiki/Open_Source_Ecology/
- Forest Farm Homestead: The Good Life Center, Harborside, Maine, http://www.goodlife.org.
- Nearing, Scott and Helen Nearing. *The Good Life: Helen and Scott Nearing's Sixty Years of Self-Sufficient Living*. New York: Schocken Books, reprint edition, 1990.
- Underground shelters: Atlas Survival Shelters, http://atlassurvivalshelters.com.
- Zero Net Energy Homes, http://www.zerohomes.org.
- Zero-Energy-Ready Homes: US Department of Energy, Office of Energy Efficiency and Renewable Energy, http://energy.gov/eere/buildings/zero-energy-ready-home.
- Eco-Friendly Shelters, http://www.ecofriendlyshelters.org.
- Hand-built Earth Homes: California Institute of Earth Art and Architecture, Hesperia, CA, http://calearth.org.

> *"The meaning of the word 'shelter' includes privacy."*
> –Frank Lloyd Wright

Off-Grid Energy

The electric grid was a great invention, and remains so as long as civilization holds together—except when it's powered by coal, oil, gas, or nuclear, which is most of the time. But grid failures are increasing, and dependence on the grid for power is a growing liability. A reasonable short-term strategy is solar power tied into the grid—so you pay the utility little or nothing but can draw on it at night in exchange for feeding into it when the sun is shining. In the long run, though, free-standing solar or wind (or other non-fossil) power sources will be essential for anyone hoping to be able to read books, play recorded music, cook dinner, or take a hot shower after sunset during the dystopian years.

- "Generating Off-Grid Power: The Four Best Ways," http://www.treehugger.com/sustainable-product-design/generating-off-grid-power-the-four-best-ways.ht.
- Real Goods Off Grid Living. http://realgoods.com and www.solarliving.com.

- Brown, Lester R. *The Great Transition: Shifting From Fossil Fuels to Solar and Wind Energy*. New York: W. W. Norton & Company, 2015.
- Rocky Mountain Institute: http://www.rmi.org.
- Ayres, Robert U. and Edward. *Crossing the Energy Divide: Moving from Fossil-Fuel Dependence to a Clean Energy Future*. Upper Saddle River, NJ: Prentice Hall, 2010.
- Solar recharging kits, http://www.goalzero.com.
- Zero Net Energy Homes, http://www.zerohomes.org.
- Zero-Energy-Ready Homes: US Department of Energy, Office of Energy Efficiency and Renewable Energy, http://energy.gov/eere/buildings/zero-energy-ready-home.
- Off-Grid Solar, http://www.backwoodssolar.com.

> *"The meta-metatrend in energy is, for lack of a better term, decentralization. Systems that were once composed of a few big technologies and a few big companies—along with thousands or millions of passive consumers—are beginning to be replaced by recombinant swarms of small producers and consumers engaged in millions of peer-to-peer transactions with a wild and woolly mix of small-scale technologies."*
>
> –David Roberts, *Grist* magazine

Off-Web Communications

In the digital age, communications that are free of hacking, intrusive surveillance, scams, ID theft, etc., are increasingly difficult to secure—yet will be essential to survival when the Internet breaks down, power outages spread, and cybercrime cripples the institutions we have relied on for almost everything.

The bad news is that we are almost totally dependent on satellite signals and the Internet—and therefore enormously vulnerable to intelligence and information blackouts. The good news is that every step taken to disengage from unnecessary dependence on the digital web is a step toward a liberated future. The most fundamental step is replacing remote, electronic communications with face-to-face, wherever possible. That, in turn, is facilitated by replacing most long-distance contacts with local ones. If you live in Texas and depend on a disembodied voice in India for tech support, the day may come when you find yourself in sudden silence—cut off.

- Ortutay, Barbara. "Disconnecting in a Too-Connected World," NBC News.com, http://www.nbcnews.com/id/43170929/ns/travel-season_travel/t/disconnecting-too-connected-world.
- Johnson, Chandra. "Face Time vs. Screen Time: the Technological Impact on Communications," *Deseret News*, August 29, 2014.

- Wolpert, Susan. "In Our Digital World, Are Young People Losing the Ability to Read Emotions?," UCLA Newsroom, http://newsroom.ucla.edu/releases/in-our-digital-world-are-young-people-losing-the-ability-to-read-emotions?
- Carr, Nicholas. "Is Google Making Us Stupid?," *The Atlantic*, July/August, 2008.
- Maushart, Susan. "Could Your Family Unplug from Technology for Six Months?," http://www.today.com/id/today-today_books/t/could-your-family-unplug-technology-six-months.

> *"It's just when people are all engaged in snooping on themselves and one another, that they become anesthetized to the whole process. Tranquilizers and anesthetics, private and corporate, become the biggest business in the world just as the world is attempting to maximize every alert. . . . The more data banks record about each one of us, the less we exist."*
> –Marshall McLuhan

Light-Footprint Mobility

Right now, in the second decade of the twenty-first century, 99.9 percent of human mobility is mechanized and largely dependent on fossil fuels. In the post-apocalyptic future envisioned by the survival protectorate, the most important forms of mobility for intelligent people will likely be walking or running, followed by the use of bicycles, scooters, and rugged solar-electric vehicles. Trucks and ships may be powered by hydrogen fuel cells. But more localized economies and closer-to-home work and resources will enable the bulk of commuting to become non-motorized. Phasing out petroleum-fueled cars and aircraft will do wonders for the health of both humans and the planet.

- League of American Bicyclists, http://bikeleague.org.
- Rails-to-Trails Conservancy, http://www.railstotrails.org.
- SLoCaT: Partnership on Sustainable Low-Carbon Transport, http://www.slocat.net.
- Road Runners Club of America, http://www.rrca.org.
- International Mountain Bicycling Association, http://www.mba.com.
- American Hiking Society, http://americanhiking.org.

> *"By the late '80s I'd discovered folding bikes, and as my work and curiosity took me to various parts of the world, I usually took one along. . . . I felt more connected to the life on the streets than I would have inside a car or in some form of public transport.*
> –David Byrne, *Bicycle Diaries*

Light-Footprint Spirituality

Including this category in the Appendix will doubtless cause some tough-guy survivalists to think I've gone too "soft" to know what it takes to get through WTSHTF alive. I think the opposite is true. A near-sighted preoccupation with having the right weapons, vehicles, and foraging skills—without also having a strong sense of how humanity will need to outgrow its fear-driven politics and institutionalized violence—would only guarantee that what happened in the Holocaust, Stalin's Russia, Hiroshima, the My Lai Massacre, the Rwanda Genocide, 9-11, the ISIS beheadings, and the murders of nine bible-study worshipers in South Carolina will continue to happen in the times to come.

The challenge now is to go forward with great inspiration and *spiritual strength*, but without the arrogance and fanaticism of religious doctrine that enabled the horrors of the Crusades, Inquisition, and a thousand other religious slaughters of civilization's past. There's reason to believe that the human spirit, empowered by love and hope, rather than fear, is the most powerful guidance system ever known. And the brainiacs of Silicon Valley didn't invent it—evolution did. If we want to reinvent our future in a way that frees us from our destructive past, we need to let that spiritual power regenerate.

- American Humanist Association: "Spiritual but not religious: a humanist perspective, http://americanhumanist.org/.
- Wilson, Edward O. *Biophilia*. Cambridge, MA: Harvard University Press, 1986.
- Unitarian Universalist Association, http://www.uua.org.
- "The Quaker Experience": *Friends Journal*, http://www.friendsjournal.org.
- Global Nonviolent Action Database, http://nvdatabase.swarthmore.edu.
- Religious Tolerance, www.religioustolerance.org.
- Union Theological Seminary (nondenominational Christian), https://utsnyc.edu/.
- Shambhala Mountain Center, 151 Shambhala Away, Red Feather Lakes, CO 80545, http://www.shambhalamountain.org.
- Spirit Rock Meditation Center, P.O. Box 169, Woodacre, CA 94973, http://www.spiritrock.org/.

> *"To not give credence to the spiritual life is to deny the validity of human experience."*
> –Albert Einstein

Safekeeping of Knowledge

Centuries ago, people developed technologies of knowledge preservation that worked for centuries. What they recorded is as retrievable from those devices today as when they recorded it. Now, you may have written the Great American Novel on an Apple computer you can't retrieve it from without paying for a specialized service that itself may not be available in another ten years when the market for such retrieval is dead. Even as human knowledge has multiplied vastly, our capacity for long-run safekeeping has shrunk.

So, we face two related challenges: The technologies of knowledge storage are plagued by obsolescence, driven by the lucrative economics of new techs replacing "old," and our nearly total dependence on soon-to-be-obsolescent techs makes us increasingly vulnerable to being stranded without access to essential information if we can't keep up with the frantic pace of technological turnover—or if we consider such keeping-up a great misuse of our time. Whatever we choose to entrust with the preservation of knowledge for the human future should ideally be portable, durable, and obsolescence-proof. For a while, the most trustworthy devices may be traditional books, maps, and personal journals—and maybe one electronic device such as a library-in-a-solar-powered-Kindle, with a couple of same-model extras for backup.

- Birkerts, Sven. *The Guttenberg Elegies: The Fate of Reading in an Electronic Age*. New York: Faber & Faber, 2006.
- Project Guttenberg: over 49,000 free e-books including Melville, James Joyce, Oscar Wilde, Jane Austin, Mark Twain, and others with copyrights expired, http://www.guttenberg.org.
- Free ebooks.net, http://www.free-ebooks.net.
- *Fahrenheit 451*, Ray Bradbury, Simon & Schuster, http://www.books.simonandschuster.com/Fahrenheit/451/.
- Solar charging kits for e-book archives, http://www.goalzero.com/portable-solar-charger.
- Indigenous Knowledge, http://www.bioneers.org/indigenous-program.

> *"History, despite its wrenching pain, cannot be unlived, but if faced with courage need not be lived again."*
> –Maya Angelou

Freedom from Tech-Addiction

For many, it has only recently begun to dawn that digital devices, games, and entertainments can be as addictive as heroin. And if the experience of drug addicts, alcoholics, and chain smokers is any indication, a majority of tech addicts may never get free. That prospect is worsened by the reality that even for those who aren't physiologically addicted, the sheer *dependence* on techs may by now be almost impossible to get free of.

But there's hope to be found in the fact that at least some people—a determined, strong, and in some ways lucky minority—have been able to fully free themselves from the demons of the pre-digital addictions. Many former chain smokers, for example, have fully overcome cigarettes by taking up running. "Positive" addictions—to aerobic sports, hiking, reading—can replace self-destructive ones.

- Carr, Nicholas. *The Shallows: What the Internet is Doing to Our Brains.* New York: W.W. Norton & Company, 2011.
- Hall, Alena. "Technology is Taking over the Most Human of Jobs, and I am Not OK with It," *Huffington Post,* June 3, 2015.
- Yeomans, Matthew. "Are Children Consuming Too Much Digital Technology?," *The Guardian,* June 18, 2013, http://www.theguardian.comg/sustainable-business/children-consuming-too-much-digital-technology.
- McVeigh, Tracy. "Internet Addiction Even Worries Silicon Valley," *The Guardian,* July 29, 2012, www.theguardian.com/technology/2012/july/29/internet/addiction-hooked-digital-technology.
- Digital Responsibility, http://www.digitalresponsibility.org/technology-addiction-awareness-scholarship.

> *"It was one thing to use computers as a tool, quite another to let them do your thinking for you."*
> –Tom Clancy, *The Hunt for the Red October*

Conscious Communities . . . and Country without Empire

Civilization has been a succession of empires, and empires are unsustainable. Just as 99 percent of all the species that have ever lived on earth are now extinct, so are 99 percent of all the empires. The American empire may be—or at this stage may already have been—the last.

Less hegemonic entities that have distinctive combinations of regional environment, climate, culture, and luck, and that don't have designs on the resources or control of other regions, may have greater sustainability but still have great vulnerabilities. Survival has always been as much a matter of social intelligence and vision, as of military force and propaganda. Strong, conscious communities may survive through the coming time when nations cannot.

- Project Tristar, http://www.projecttristar.net.
- Transition Towns: https://en.wikipedia.org/wiki/Transistion_town
- Combating Globalization, http://www.combatingglobalization.com.
- Earth Day Network, http://www.earthday.org.
- Fellowship for Intentional Community, http://www.ic.org.
- Life in Cooperative Culture: *Communities* magazine, http://www.ic.org/Communities-magazine.
- Dancing Rabbit Ecovillage (Missouri), http://www.dancingrabbit.org.
- Earthaven Ecovillage (North Carolina), http://www.earthhaven.org.
- Los Angeles Ecovillage, http://www.laecovillage.org.
- Sustainable Communities Network, http://www.sustainable.org.
- Wilson, Reid. "Which of the 11 American Nations Do You Live In?," *The Washington Post*, November 8, 2013, http://www.washingtonpost.com/blogs/gtovbeat/2013/11/08/Which-of-the-11-American-nations-do-you-live-in.
- Pinker, Steven. *The Better Angels of Our Nature: Why Violence Has Declined*. New York: Viking Adult, 2011.

> *"Human beings will be happier, not when they cure cancer or eliminate racial prejudice or flush Lake Erie, but when they find ways to inhabit primitive communities again. That's my utopia."*
> –Kurt Vonnegut

Post-Global Economy

Hundreds of pop-economic "consultants" have warned of impending economic collapse, often predicting specific dates—thereby spurring the sales of seminar tickets or books. So far, the dates of predicted collapse have always passed without the slightest tremble in the global edifice. In comparison, very few serious economists (or interdisciplinary scholars) have addressed the question of how human economies will work after the current *civilization* fails. The pop-economic savants get it wrong because they fail to see the money economy as inseparable from the environment, climate, diminishing natural capital, and culture. A few far-seeing authors and groups are now tackling that larger question—asking how humanity can build a post-dystopian economy that does not further deplete the natural resources of the planet and does not depend on the myth of a perpetually growing GDP.

- "Overcoming Consumerism" Verdant, http://www.verdant.net.
- Redefining Progress, http://progress.org.
- *Journal of Industrial Ecology*, http://jie.yale.edu.
- Center for the Advancement of the Steady-State Economy, http://steadystate.org/category/herman-daly.
- Schumacher, E. F. *Small is Beautiful: Economics as if People Mattered*. New York: Harper Perennial, 1989.

- Anti-Consumerism and the Alternate Economy, http://www.verdant/alternate_economy.htm.
- Ayres, Robert U. *The Bubble Economy: Is Sustainable Growth Still Possible?* Cambridge, MA: The MIT Press, 2014.

> *"Perpetual Economic Growth is neither possible nor desirable. Growth, especially in wealthy nations, is already causing more problems than it solves. The positive, sustainable alternative is a steady-state economy."*
> –Herman Daly
> Professor, University of Maryland School of Public Policy

Envisioning the Future

Imagining how the human prospect might change has been a preoccupation of intelligent individuals ever since it became evident that societies *can* change. That interest has historically gone hand-in-hand with the development of technology. In the past century of highly accelerated invention and technological revolution, the imagining has shifted from the work of ingenious individuals (e.g., Leonardo da Vinci sketching flying machines over four centuries before they were successfully built), to research organizations engaged in a race to save humanity from destruction. Some of these groups are still guided by a techno-optimist bias; others are more insightful about the realization that the human future can't be separated from its interdependence with the natural environment, on which we depend for every step or breath we take.

- Union of Concerned Scientists, http://iww.ucsusa.org.
- Co-Intelligence Institute: http://co-intelligence.org
- The Future of Humanity Institute, University of Oxford, www.fhi.ox.ac.uk.
- The Future of Human Evolution, http://futurehumanevolution.com.
- World Future Society, http://www.wfs.org.
- Ehrlich, Paul R., and Ornstein, Robert E. *New World New Mind: Moving Toward Conscious Evolution.* Malor Books, 2000.
- Future Conscience, http://www.futureconscience.com.
- The Millennium Alliance for Humanity and the Biosphere, https://mahb.stanford.edu/.
- Future Timeline, http://www.futuretimeline.net.
- Brown, Lester R. *The Great Transition: Shifting from Fossil Fuels to Solar and Wind Energy.* New York: W. W. Norton & Company, 2015.
- Center for a New American Dream, http://www.newdream.org.
- Bioneers, http://www.bioneers.org.

> *"It is not in the stars to hold our destiny but in ourselves."*
> –William Shakespeare, *Julius Caesar*

Enlightened Education

In all the attention that has been paid to human survival by at least that growing minority of people who sense calamity ahead (the global majority remaining to varying degrees oblivious), relatively little of that attention has been given to the nature of education. A lot of concern has been with how "I" or "my family" or "my group and I" will survive—what weapons, vehicles, foods, etc., we will need. Little has been with the education of the children who will have to carry on after we are gone.

Yet, fundamental failures in education are at the root of the collapse we now face. Worldwide, the bulk of education has been authoritarian—based on unchallenged religious or ideological doctrines. It has been based on *established answers* to every form of human questioning. But overwhelming evidence suggests that those answers have been tragically flawed, and the kinds of education that can serve a sustainable human future will need to be based on *new questions*, perhaps of very different kinds than the ones to which our teachers thought they had answers.

- Bioneers Education For Sustainability, http://www.bioneers.org.
- Truthout, http://www.truth-out.org.
- Gaia Education, http://www.gaiaeducation.net.
- Thom Hartmann Program (streaming live radio, video, and television on the web), http://www.thomhartmann.com.
- *EcoIQ Magazine*, www.ecoiq.org.
- Thilman, James. "'Most Likely to Succeed': Schools Should Teach Kids to Think, Not Memorize," *Huffington Post*, http://www.huffingtonpost.com/2015/04/24/most-likely-to-succeed.
- Teaching and Learning for a Sustainable Future, United Nations Educational, Scientific, and Cultural Organization (UNESCO), http://www.unesco.org/education/tlsf/.
- Kennedy, John F. "The Soft American," *Sports Illustrated*, December 26, 1960.
- Institute for Ethics and Emerging Technologies, http://ieet.org.
- Progressive Education, http://progressiveeducationinstitute.org.
- Ayres, Ed. "Why Are We Not Astonished?," *World Watch*, May/June 1999, http://www.worldwatch.org/sysstem/files/EP1238.pdf.

"When you stop learning, stop listening, stop looking and asking questions, always new questions, then it's time to die."
–Lillian Smith

"A question is the most powerful force in the universe."
–TV ad for Google

Emergency Preparedness

Unfortunately, for too many people, waking up to the reality of an endangered civilization will only happen when they get hit personally, head-on, by hurricanes, floods, tornadoes, wildfires, or—for farmers—crop-destroying drought. All of those events are becoming more frequent, now, as other direct-hit disasters such as terrorist attacks or infrastructure failures may soon be. But these events will also be, for those who are adept at learning, perceptual gateways to the larger threat of cascading failures of whole systems (power grid, Internet, municipal budgets, banking, and so on) leading inexorably to a failing of civilization at large. Emergency preparedness is a useful first step in preparing for the longer term.

- Centers for Disease Control, http://www.emergency.cdc.gov.
- American Red Cross, http://www.redcross.org.
- International Federation of Red Cross and Red Crescent Societies, http://www.ifrc.org.
- Federal Emergency Management Agency, http://www.fema.gov.
- Off Grid Survival, http:/offgridsurvival.com.

> *"There's no harm in hoping for the best, as long as you're prepared for the worst."*
> –Stephen King, *Different Seasons*

The Warnings of Science

The first real warnings—*not* religious prophecies or post-apocalyptic sci-fi movies—that civilization is failing began, significantly, around the beginning of the industrial revolution, when technology expanded human powers vastly, while the wisdom of those who wielded those technologies arguably did not expand at all. The intelligence of inventors and investors may have been fatally numbed, in fact, by the temptations of quick profit and wealth for some, and quick gratification and addiction for the rest. The warnings may be said to have begun with Thomas Malthus, later updated by a host of iconic scientists in the twentieth and early twenty-first centuries.

- *World Scientists' Warning to Humanity*, Union of Concerned Scientists, 1992, wwwucsusa.org.
- Union of Concerned Scientists, www.uscusa.org.
- Ringerberg, Lynn. "A Dramatic Doomsday Warning to the World," *CNN*, January 23, 2015, http://www.cnn.com/2015/01/23/opinion/ringenberg-doomsday-clock.

- Schlosser, Eric. *Command and Control: Nuclear Weapons, the Damascus Accident, and the Illusion of Safety*. New York: Penguin, 2014.
- Intergovernmental Panel on Climate Change (IPCC), Fifth Assessment Report, www.ipcc.ch/report/ar5.
- "Journey to the Earth," PBS video, http://video.pbs.org/program/journey-planet-earth/.
- Unsustainable Population Growth: Population Matters, www.populationmatters.org.
- Ehrlich, Paul R., and Anne H. Ehrlich, *The Population Explosion*. New York: Frederick Muller, 1990.
- Brown, Lester R. *The 29th Day: Accommodating Human Needs and Numbers to the Earth's Resources*. New York: W. W. Norton & Company, 1978.

> *"We are on a collision course to disaster if we continue to damage the planet"*
> –World Scientists' Warning to Humanity

Index

CPSIA information can be obtained
at www.ICGtesting.com
Printed in the USA
FFOW03n1255041116
29090FF